# Urban Water Demand Management and Planning

# Urban Water Demand Management and Planning

**Duane D. Baumann**

**John J. Boland**

**W. Michael Hanemann**

**With a Foreword by Gilbert F. White**

**McGraw-Hill, Inc.**

New York  San Francisco  Washington, D.C.  Auckland  Bogotá
Caracas  Lisbon  London  Madrid  Mexico City  Milan
Montreal  New Delhi  San Juan  Singapore
Sydney  Tokyo  Toronto

# McGraw-Hill

### A Division of The **McGraw·Hill** Companies

Library of Congress Cataloging-in-Publication Data

Baumann, Duane D.
    Urban water demand management and planning / Duane D. Baumann,
John J. Boland, W. Michael Hanemann.
        p.     cm.
    Includes index.
    ISBN 0-07-050301-X
    1. Municipal water supply—Management.    2. Water conservation.
I. Boland, John J.    II. Hanemann, W. Michael.    III. Title.
TD353.B36    1997
333.91'2—dc21                                                        97-30667
                                                                        CIP

1 2 3 4 5 6 7 8 9 0    FGR/FGR    9 0 2 1 0 9 8 7

The sponsoring editor for this book was Larry Hager, the editing
supervisor was Bernard Onken, and the production supervisor was
Pamela A. Pelton. It was set in Palatino by North Market Street Graphics.

Printed and bound by Quebecor/Fairfield

This book was printed on recycled, acid-free paper containing a
minimum of 50% recycled de-inked fiber.

# Contents

Foreword     xi
Acknowledgments     xiii


## 1. The Case For Managing Urban Water     1

The Urban Water Industry     1
The Evolution of Urban Water Use     2
The Economic Importance of the Urban Water Industry     4
Historical Emphasis on Supply Alternatives     6
Water Conservation: Renewed Prominence in the 1980s     7
What Is Conservation?     9
Measurement of Conservation Effectiveness     17
Integrating Demand and Supply Alternatives     18
Purpose of Volume     20
Water Management Myths     21
  Water Is a Necessity     21
  The Myth of Per Capita Use     22
  Water Users Do Not Respond to Price     27
  The Myth of Excessive Irrigation of Urban Landscapes     27
  Conservation Will Lead to Negative Fiscal Impacts on
    the Water Supply Agency     28


## 2. Determinants of Urban Water Use     31

Industrial Water Use     33
  Economic Theory of Demand for Water as an Input     34
  Implications for Empirical Modeling of Industrial Water Use     47

v

Residential Water Use                                                  52
  Economic Theory of Consumer Demand for Water       56
  Empirical Estimates of M&I Demand Elasticities          65

## 3. Forecasting Urban Water Use: Theory and Principles    77

Need for Water Use Forecasts                               77
  Long-Range Planning                                      78
  Supply Adequacy Evaluation                               79
  Short-Range Planning                                     80
  Financial Planning                                       80
Criteria for Selecting a Forecast Method                   81
  Principles of Forecasting                                81
  Distinction Between the Forecast and the Method          82
  Selection Criteria                                       82
  Correlates of Accuracy in Water Use Forecasts            83
Evaluation of Water Use Forecast Methods                   84
  Time Extrapolation                                       85
  Bivariate Models                                         85
  Multivariate Models                                      87
Uncertainty in Forecasts                                   89
  "Safety Factors"                                         89
  Scenario Approaches                                      90
  Sensitivity Analysis                                     91
  Contingency Trees                                        91
  Forecast Bounds                                          92
Forecasting Under Uncertainty                              92
  Base Forecasts                                           92
  Expressing Uncertainty                                   93
  Interpretation of Results                                93

## 4. Forecasting Urban Water Use: Models and Applications    95

IWR-MAIN Water Demand Analysis Software                    95
  Background                                               95
  Forecasting Methods                                      99
  Conservation Savings Methods                            103
  Benefit/Cost Analysis Methods                           106
  Summary                                                 108
Water Requirements Modeling with the Installation Water
Resources Analysis and Planning System (IWRAPS©)          108
  Building Types and Sizes                                111
  Construction and Demolition                             112
  Seasonality                                             112
  Climate                                                 113
  Installation Mission                                    113
  Installation and Fixed Effects                          114
  Conservation and Mobilization                           115
  Water-Requirement Algorithms                            115

Sectoral Allocation 120
Conservation Algorithm 122
Forecast Procedures 123

**5. Price and Rate Structures** **137**
Components of a Water Rate Structure 137
Criteria for Designing Water Rates 139
Criteria for Revenue Generation 139
Criteria for Cost Allocation 140
Criteria for Providing Incentives 140
Applying the Criteria in Practice 141
Water Supply Costs and Complexities 143
Marginal Cost Pricing 148
The Economic Argument for Marginal Cost Pricing 149
Seasonal Rates 158
Increasing Block Rates 165

**6. Forms and Functions of Water Pricing: An Overview** **181**
What Is the Price of Water? 181
Types of Decisions Affected by Water Price 182
The Conflicting Roles of Water Price 183
Pricing and Water Quality 185
Pricing and Water Supply Reliability 186
Conclusions 187

**7. Phoenix Changes Water Rates from Increasing Blocks to Uniform Price** **191**
Information About the Phoenix Water System 192
Genesis of Interest in Changing Water Rate Structures 192
The Previous Water Rate Structure 194
Study of Alternative Water Rate Structures 195
The Current Water Rate Structure 200
Impacts of Current Rate Structure on Customer Classes and Selected Customers 201
Summary of Current Rate Structure Benefits 201
An Attempt to Change the Water Rate Structure Again 202
Concerns with the Current Rate Structure 203
Customers Below Lifeline Cannot Impact Bill by Conserving 203
Customers with Nonseasonal Demand Experience Lower Summer Bills 203
The Effectiveness of the Current Lifeline in Assisting Low-Income Customers 204
The Negative Impact of the Lifeline on Large Water Use Customers 205

Water Services Department Concerns and Recommendations     206
  Cost-of-Service Update                                    206
  Equal Lifeline in All Months or Seasons                  206
  Meter Maintenance Costs in the Volume Charge
  (Equal Service Charge for All Meter Sizes)               207
Rate Structure Options                                       207
Discussion of the Rate Options in Respect to Rate Objectives  208
  Revenue Sufficiency                                      209
  Equity                                                   210
  Efficiency                                               210
  Social Acceptability                                     211
  Practical Feasibility                                    212
  Water Conservation                                       212
Customer Bill Impacts                                        212
  Single-family Customers                                  213
  Multifamily Customers                                    214
  Commercial Customers                                     216
  Irrigation Meters                                        216
  Summary of Customer Impacts                              216
Summary Analysis of Proposed Rate Structures                 217
Reaction to Recommended Rate Structure                       218
Conclusion                                                   219

**8. Trends in Revenues and Expenditures for Water and Sewer Services: Implications for Demand Management**                                              **221**

Categories of Expenditures                                   222
  Trends in Revenues and Expenditures                      223
  Aggregate Trends                                         225
  Household Trends                                         227
  Debt                                                     230
Relationships in Self-Financed Utilities                     231
Discussion of Trends                                         233
  Water Use                                                233
  Operating Costs                                          233
  Influence of Financing Practices                         234
Summary and Conclusions                                      235

**9. Demand Management Planning Methods**                        **237**

Establish Program Goals                                      238
Determine Applicability and Feasibility                      239
  A Library of Conservation Measures                       239
  Screening Tests                                          240
Determine Social Acceptability                               240
  Review Objectives and Define Target Population            243
  Define Data to Be Collected                              244
  Select a Method of Data Collection                       245

Design and Implement Sampling Plan                                    247
Design and Implement Survey Questionnaires                           249
Data Compilation and Analysis                                         250
Estimate Potential Water Savings                                      250
Mechanical Estimates                                                  250
Empirical Estimation                                                  257
Candidate Measures                                                    257
Define Implementation Conditions                                      258
Program Contents                                                      258
Definition of Target Population and Program Participants              260
Program Incentives                                                    261
Customer Contact Modes                                                261
Schedule of Program Implementation and Duration                      262
Specification of Responsible Agencies                                 262
Program Evaluation Plan                                               262
Conduct Benefit-Cost Analysis                                         263
Demand Management Costs                                               263
Demand Management Benefits                                            267
The Accounting Perspective                                            268
Accounting for Intangibles                                            269
The Process of Discounting                                            271
Benefit-Cost Measures                                                 272
Integrate Water Conservation into Water Supply Plans                  274
Proposal Development Principles                                       274
Development of Alternative Conservation Proposals                     276
Supply Reliability Considerations                                     278
Documentation of Water Management Plans                               279

## 10. Demand Management Program Evaluation Methods

**10. Demand Management Program Evaluation
Methods**                                                           **283**

The Role of Program Evaluation                                        283
Objectives of Program Evaluation                                      285
Process Evaluation versus Impact Evaluation                          286
Data Collection for Process Evaluation                                288
Data Collection for Impact Evaluation                                288
Sample Selection for Evaluation                                       289
Design of Program Evaluation Studies                                  290
Validity of Evaluation Studies                                        291
Measuring Program Impact                                              292
Data and Methods for Measuring Program Impacts                        293
Using Engineering Estimates                                           294
Using Statistical Comparisons                                         294
Using Multivariate Regression                                         296
Combined Evaluation Methods                                           298
Uncertainty in Measuring Water Savings                                298
Standardizing Water Savings Estimates                                 299
Long-Term Monitoring                                                  300

**11. Integrating Water Supply and Water Demand
Management**                                                          **303**

    Integrated Resource Planning (IRP)                               304
    The IRP Process                                                  309
    Case Studies in Water IRP                                        310
      New York City                                                311
      Wichita, Kansas                                              314
      Seattle, Washington                                          316
      Southern Nevada Water Authority                              319
      Metropolitan Water District of Southern California           321
    Conclusion                                                       324

**12. Application of Integrated Resource Planning
Approach to Urban Drought**                                          **329**

    Droughts and Water Management                                    330
    Long-Term Protection Against Droughts                            330
      Short-Term Drought Management                                332
      Proactive Drought Management                                 333
    A Drought Planning Framework: DROPS                              333
      The DROPS Approach                                           333
      The Coping-Cost Criterion                                    336
    Example of Phoenix, Arizona                                      336
      Balance of Supply and Demand                                 337
      Potential Shortages                                          338
      Options for Dealing with Drought                             339
      Long-Term Drought Protection Alternatives                    340
      Analysis of Drought Protection Trade-offs                    341
    Summary                                                          344

    Index    345

# Foreword

In a time when modes of water management around the world are undergoing fundamental change, this appraisal of methods of dealing with urban water demand meets an immediate, urgent need.

The changes in social and engineering approaches to human use of water are related to the convergence of several trends. Many of the cheapest sites for water storage and conveyance have been developed. The full impacts of such developments—post and future—on environmental systems are being recognized more accurately. Social criteria for evaluating development are becoming more refined and the role of local watershed groups is enlarging.

For these and other reasons it is increasingly important to cultivate precision and reliability in determining demand, forecasting demand, and evaluating the various economic, technologic, and social determinants of demand for water. With the growing significance of the urban sector, as demonstrated in megacities and in sprawl into desert areas, the recent experience in the United States deserves thoughtful appraisal. The old paradigm of designing the cheapest reliable supply with little attention to demand determinants, pricing structures, and financing policies is no longer suitable.

This book provides critical reviews of new and emerging methods and shows how they may be incorporated into integrated resource planning, including supply management, and how their effectiveness in practice has been and may be evaluated. The analytical frameworks are complemented by a realistic appraisal of how efforts at demand management fare in the daily political and administrative life of a growing city.

A basic task inherent in the whole effort is to define the dimensions and implications of the concept of water conservation. As urban societies strive for genuine sustainability, it is essential to recognize realistically what wise demand management requires and avoids.

*Gilbert F. White*
*University of Colorado*

# Acknowledgments

This volume is about advanced methods for urban water-demand analysis and management. The topics and authors have been chosen to cover a range of topics, and to present both principles and details of application. In many cases, the material provided here is state of the art, some of it previously unpublished. Yet the topic and the authors owe a considerable debt to the individuals and institutions that pioneered water demand management in the United States.

The basic foundation for this perspective on water management was laid down in the 1950s at the University of Chicago and Harvard University. At Chicago, Gilbert F. White argued persuasively for a broadening of the range of choice considered in water resources planning. His early work on flood hazard broke with the traditional approach of reliance on structural measures, demonstrating the considerable advantages of reducing potential damage through floodplain management, and mitigating damage through national insurance programs. His students, such as Robert W. Kates at Clark University, extended the paradigm into urban water resources management, especially in the path-breaking work on urban drought with Clifford Russell (then at Harvard) and David Arey (then at Clark).

At Harvard, a different but fully consistent line of investigation was begun. From 1956 until the early 1960s, the Harvard Water Program, directed by Arthur Mass, firmly established the roles of demand analysis, economic tools, and optimization in the planning and management of urban water systems. Mass assembled a truly multidisciplinary team, including Robert Dorfman, Otto Eckstein, Maynard Hofschmidt, John

Krutilla, Stephen A. Marglin, Gordon Maskew, and Harold A. Thomas, Jr. The results of their efforts had a profound and lasting effect on generations of researchers and practitioners.

Similarly at the same time the Rand Corporation supported the seminal work of Jack Hirshleifer, James C. De Haven, and Jerome W. Milliman. Their work focused on the importance of alternatives that promoted efficiency and economy in urban water. The groundwork was laid calling for the need for demand analyses in efficient urban water resources planning and management.

Concurrently, at the Johns Hopkins University, John Geyer, F. Pierce Linaweaver, Jerome B. Wolff, and others completed the first comprehensive analyses of residential, commercial, and institutional water use. Later extensions of the work by Linaweaver and Charles W. Howe (then at Resources for the Future) produced the seminal Howe and Linaweaver models of residential water use, still widely applied after thirty years.

Urban water-demand research was reinvigorated in the later 1970s when the U.S. Army Corps of Engineers Institute for Water Resources (IWR) launched a research and development program in the area. Originally intended to support water planning activities within the Corps, the IWR effort soon began to generate analyses, tools, and models with wide application in the water field. This work continued for nearly a decade, directed and/or inspired by Corps staff members such as James Crews, Donald Duncan, Michael Krause, Darrell Nolton, and Kyle Schilling. Planning and Management Consultants, Ltd. (PMCL), was closely connected with this work, and much of the success in the development and testing of new tools can be attributed to Benedykt Dziegielewski, Jack Langowski, and Eva Opitz.

The purpose of all this effort, of course, has been to improve the effectiveness and efficiency of urban water management and planning. But the line dividing research and the first stages of practical application is sometimes vague and hard to define. The early applications of these approaches and models required uncommon skill and courage, and their success is a tribute to the people involved. Among the most skilled and dedicated of the early adopters are Wiley Horne of the Metropolitan Water District of Southern California and William Mee of the Phoenix Water and Wastewater Department. Within the Corps of Engineers, early and often innovative applications were due to the efforts of Eugene Stakhiv, William Werick, and Germaine Hofbauer. Finally guidance for urban-demand analyses was promoted through support by Lyle Hoag, then Executive Director of the California Urban Water Agencies, in preparation of the handbook on water conservation subsequently published by the American Water Works Association.

**Table 1-1.** Urban Water-Use Observations
in Gallons Per Capita Per Day (GPCD)

| Year | City | GPCD |
|---|---|---|
| 97 | Rome | 38 |
| 1550 | Paris | 0.25 |
| 1885 | Philadelphia | 72 |
| 1890 | Paris | 65 |
| 1895 | Philadelphia | 162 |
| | Baltimore | 95 |
| 1900 | "National average" | 90 |
| | "Typical city" | 100 |
| | "Major industrial city" | 159 |
| 1913 | United States | |
| | Dallas | 56 |
| | New York | 129 |
| | Chicago | 275 |
| 1913 | Europe | |
| | Vienna | 14 |
| | London | 40 |
| | Paris | 98 |
| 1940 | U.S. survey—mean | 127 |
| 1954 | U.S. survey—mean | 140 |
| 1965 | U.S. survey—mean | 156 |
| 1970 | U.S. survey—mean | 189 |
| 1981 | U.S. survey (sample size 137) | |
| | Maximum | 493 |
| | Mean | 176 |
| | Minimum | 86 |
| 1989–92 | U.S. survey—mean | 180 |

SOURCES: Frontinus, 1899; Sewerage Commission, 1897; Babbit and Doland, 1929; AWWA, (undated); AWWA, 1981; AWWA, 1992.

expanding to meet the needs of the population, water use levels rose dramatically. Table 1-1 presents data from various sources to illustrate this trend as well as the large differences among cities that have been observed and continue to be observed.

The surprisingly large water use for imperial Rome, as calculated from Frontinus' data by Clemens Herschel in 1899, reflects relatively few direct connections to buildings, but a substantial water use at public fountains and baths. On the other hand, Paris in the sixteenth century evidently lacked such public amenities. Late-nineteenth-century estimates show the effect of rapid growth in the number and intensity of urban water uses. Some cities, such as Chicago in 1913, show very high per capita use. It should be noted that Chicago's water supply was unmetered at that time.

Table 1-1 shows mean per capita water use for U.S. cities from 1940 onward, based on periodic surveys conducted by the American Water Works Association. Superficial examination of the data would suggest a strong positive growth in water use, followed by a leveling or dropping after 1970. In fact, there is little evidence for this. What the numbers show is (1) sampling errors due to the voluntary and limited nature of the survey and (2) the effect on the national mean of a significant population movement from the humid (low water use) Northeast to the semiarid (high water use) Southwest. There has been no particular time trend in urban water use during the past fifty-odd years. Differences observed from time to time and from place to place are readily explained by differences in price, income, climate, housing types, and other determinants of water use (see Chapter 2).

## The Economic Importance of the Urban Water Industry

From an economic, if not an operational, perspective, the urban water industry consists of those activities necessary to secure, treat, transport, store, and distribute water for various human uses (water supply) as well as additional activities needed to collect, transport, treat, and dispose of the resulting wastewater (wastewater disposal). Although these functions are sometimes separated organizationally, they are inseparable in other ways. With the exception of certain outdoor uses (lawn and garden irrigation, for example), each water use generates a wastewater flow. In most urban areas in the industrial world, water supply creates the need for wastewater disposal. The cost of water use is the sum of water supply and wastewater disposal costs.

In the United States, the water supply function is carried out by both public and private sector organizations. Of the 200-odd million Americans served by public systems, about one-seventh obtain their water from investor-owned water utilities. The remaining, government-owned utilities are municipal agencies, regional agencies, government corporations, or state agencies. Although there are some investor-owned wastewater disposal organizations, nearly all are government-owned utilities. In recent years, there has been increasing use of private-sector organizations to operate all or part of government-owned utilities. In some cases, ownership of the physical plant may pass to the private operator. So far, these privatization efforts have not significantly affected the public-private water supply split, which has been stable for many years.

In considering the economic importance of the urban water industry, two things can be noted: it is large in absolute terms, and it is exceptionally capital-intensive. In 1993, government-owned water supply utilities collected $20.8 billion in net revenue from all sources (U.S. Bureau of the Census 1996e). A comparable estimate for the investor-owned segment is $3.2 billion (U.S. Bureau of the Census 1996c; note that these data, the latest available, are for the prior year). Total revenue to water supply utilities, then, can be estimated at about $24 billion. During the same period, adjusted gross revenue received by government-owned wastewater utilities was $16.5 billion; revenue to private-sector wastewater utilities was $0.4 billion, for a total of $16.9 billion (U.S. Bureau of the Census 1996c and 1996e).

The industry as a whole collected about $41 billion in revenue throughout the United States in fiscal year 1993. This is a relatively small fraction of aggregate personal income for the same period—about 0.8 percent—but it is almost equal to the total output value of the mining industry in 1993 ($41.2 billion) and about half of the output value of the entire agriculture, forestry, and fishing sector of the economy ($91.0 billion) (U.S. Department of Commerce 1996).

Also during fiscal year 1993, government-owned water supply utilities invested some $6.2 billion in fixed assets (U.S. Department of Commerce 1996, p. 53). Extrapolating that amount to include the investor-owned segment gives about $7.2 billion for water supply facilities. Another $10.3 billion was invested in government-owned wastewater disposal systems (U.S. Department of Commerce 1996, p. 53; Vogan 1996). The industry total is estimated at $17.5 billion, equal to 43 percent of adjusted gross revenue for that year.

No other major industry category in the United States even approaches this ratio of annual investment to revenue. Another capital-intensive industry group, All Mineral Industries (SIC 101-149), reported 1992 investments which averaged 11 percent of revenues. The most capital-intensive individual industries in this group, such as Chemical and Fertilizer Mining (SIC 1479), reached a 26 percent investment ratio (U.S. Bureau of the Census 1996b). United States manufacturing industries (SIC 201-399) invested only 3.4 percent of revenues in 1992 (U.S. Bureau of the Census 1996a). The most capital-intensive industries, after water, are Electric Services (SIC 491) and Communications (SIC 48), which average 16 percent and 18 percent investment, respectively (U.S. Bureau of the Census 1995, 1996c, 1996d).

A further characteristic of the water industry is the very long life expected of most capital facilities. Unlike other industries, where investments may have an economic life of 5 to 20 years, facilities for water supply and wastewater disposal are almost uniformly long-lived. Few improvements are expected to last for less than 20 years and most have much longer lives. Some facilities, including pipes and dams, may function for 50 years, 100 years, or more.

All of these statistical data serve to portray an industry where the efficiency and effectiveness of capital investment is exceptionally important. This is the single most capital-intensive industry in the U.S. economy, and it doubtless holds a similar position in other countries. The size of U.S. annual investment is large ($17.5 billion in 1993, corresponding to 43 percent of total revenue) in an industry where the consequences of error are notably long-lasting (facility life of 20 to 100 years).

Since capital investments are undertaken, in part, on the basis of water use forecasts, the accuracy of those forecasts becomes a matter of serious economic concern. The way in which the costs of water supply and wastewater disposal are recovered is similarly important, since it determines both incentives for customers to demand new capital facilities and the availability of funds to finance investments. Other demand management approaches, such as long-term conservation and drought management, also assume new importance in this context. A simple hypothetical example will illustrate this point: if it is expected that a law mandating a new plumbing fixture will eventually reduce water use by an amount equal to five years' growth, and if 50 percent of capital investments are thought to be flow-related, the benefits of the new policy are equal to a one-time gain of about $44 billion (avoided investment cost for five years times 50 percent).

## Historical Emphasis on Supply Alternatives

Throughout history, concentrations of population have always been associated with large-scale water supply facilities. In the Middle East, in Central America, in the American Southwest—long-extinct civilizations have left evidence of their dependence upon water supply. The Sanitary Revolution of the late nineteenth and early twentieth centuries, by insuring the safety and palatability of urban water supplies, only increased the dependence of urban civilization on water. The dependence is no less evident today. However, as the water and other resources of the United States are more widely and intensively exploited, the efficiency with which they are used becomes of greater concern.

Unlike the past, present urban water supply planning is a drastically different, challenging, and complex task. Traditionally, the planning process started by projecting the population to be served, estimating per capita water use, and then simply multiplying one projection by the other to derive the future water use. Armed with an estimate of future water need, planners had to face the problem of identifying adequate and available sources of supply, usually additional reservoirs and/or well fields.

The problem today is not solely an inadequacy of supply; instead a wide range of factors have an influential and important role in the planning and management of our urban water resources. Consequently, new techniques of planning and methods of evaluation will need to be developed. In addition, water management policies and practices will be modified. As early as 1973, the U.S. National Water Commission noted that:

> To increase efficiency in water use and to protect and improve its quality, and to do these things at least cost and with equity to all parts of the country . . . require major changes in present water policies and programs.

Since 1973, there have been substantial changes in the process of planning and management of our urban water resources. For example, the U.S. Federal Water Resources Council developed the Principles and Standards for Planning Water and Related Land Resources; and the U.S. Army Corps of Engineers implemented new guidelines, such as the Planning Process: Multiobjective Planning Framework, and developed environmental impact analyzes for proposed projects.

Concurrent with changes in the planning process has been a shift in perspective, that is, to a broader range of alternatives. The traditional response to increasing demand for water has been the development of additional supply. Those alternatives, for example, that would modify demand have been generally ignored. Similarly, this reliance upon technologies to increase supply, instead of policies to modify the schedule of demand, has been evident in another water resources problem—flood control.

In planning for urban water, the challenge is to determine the optimum combination of all alternatives to balance the supply and demand. Not only are the alternatives that increase supply considered, but those options that modify the demand for water—such as water conservation—are also evaluated.

## Water Conservation: Renewed Prominence in the 1980s

The first indication of widespread interest in urban water conservation appeared shortly after 1970. The National Water Commission conducted a study of the potential for water use reduction through conservation practices, including pricing policy, and discussed water conservation as an alternative to, or adjunct of, water supply augmentation. Some urban water suppliers began to encourage conservation practices by their customers (for example, Washington Suburban Sanitary commission, *A Customer Handbook on Water-Saving and Wastewater-Reduction*, 1972). Further attention to water conservation grew out of the realization that reduced water use may result in reduced sewer flows. The Clean Water Act of 1977 specifically requires measures to reduce wastewater flows as a condition of eligibility for wastewater treatment facility construction grants.

Recycling, which already has a significant effect in the reduction of industrial water use, and water use efficiencies are expected to exert an increasingly important role in projected water use. Hence, the national emphasis on water conservation is already reflected in the projections: water use efficiencies are expected to exert an increasingly important role in projected water use.

However, the rationale for considering water conservation in water supply planning is not solely a function of the relationship between supply and projections of use; other factors today impinge upon the efficiency of water use and planning to meet future demand. Data on national aggregate water use have little relevance because urban water supply planning is a local phenomenon. It is at the local level where the range of factors that determine the efficiency of water supply production is of most interest.

The broadening of water planners' perspectives to include demand management alternatives and other innovative solutions has been brought about by a number of new challenges that water planners must face today and in the future.

- Untapped sources of water are becoming rarer, and the depletion and contamination of groundwater sources has further limited supplies.

- The increased frequency of droughts during the last decade has increased the competition for water between urban and agricultural interests.

- Environmental concerns about increased water use have intensified during the last two decades to the point where the development of new supplies is politically infeasible, and the prospects for financing major construction programs are discouraging for many water agencies.

- The Safe Drinking Water Act of 1974 and its recent amendments have forced many communities to comply with increasingly stringent limits on a large number of contaminants in drinking water. These new limits have significantly increased the cost of water treatment, and some water sources that had served communities for decades are no longer considered adequate because of excessive contamination.

These new considerations have forced water planners to extend their perspective beyond traditional water resource development projects. Alternative ways of increasing water supply have been explored and evaluated, including:

- More efficient utilization of existing water supplies, such as pumped storage, reduction of losses through the lining of reservoirs, and evaporation suppression and reallocation of storage water for flood protection.

- Protection of existing supplies by cleaning up the waste sources contaminating water, construction of barriers against the intrusion of salt water, and development of new technologies for large-scale treatment of contaminated aquifers.

- Development of unconventional sources of supply by using groundwater aquifers for storage of excess supply of surface water (water banking) or relying on conjunctive use of groundwater and surface water, desalinizing seawater or brackish groundwater to drinking-water quality, and reclaiming wastewater, for example, by building dual-pipe distribution systems.

- Water marketing and regional management of the existing and new sources of supply to improve the distribution of water among water users within the region.

- The use of demand management alternatives represents an important change in water supply planning. Demand reduction programs allow some agencies to balance future supply and demand at a cost that is below the economic, social, and environmental costs of new supply development.

These dynamic changes in the physical, social, and economic environment resulted in necessarily new and critical questions. It follows that answers to these questions demand the development of new methods of evaluation. Specifically, these questions were:

- What is conservation; that is, how do we define demand management?

- What is the effectiveness of available conservation or demand management measures?

- What are the principles for evaluation of water conservation measures for municipal and industrial water supply; that is, how do we formulate the optimum combination of demand and supply alternatives into an urban water resources plan.

## What Is Conservation?*

To most persons, water conservation is a noble and laudable goal, but in the formulation and implementation of water conservation policies, a formidable obstacle is immediately encountered: what exactly *is* conservation?

* This section on the definition of water conservation was published initially by the Institute for Water Resources, U.S. Army Corps of Engineers, *The Role of Conservation on Water Supply Planning,* April 1979, and in Baumann, Duane D., Boland, John J., and Sims, John of "Water Conservation: The Struggle Over Definition," *Water Resources Research,* 20 (April 1984), pp. 428–434.

Clearly, an answer to that question is needed—an explicit definition of conservation is required—before an agency can begin to formulate policy. In an attempt to fill that need, the U.S. Water Resources Council (1979) discussed water conservation as:

> Water resources planning, which fully incorporates conservation, shall be based upon systematic evaluation of alternative water resource management strategies (to include structural and nonstructural measures). General types of program, project, and policy alternatives to be fully considered separately or in combination are:
> 1. Reduce the level and/or alter the time pattern of current and future demand for selected purposes to make water available for alternative uses
> 2. Modify management of existing water developments to enhance availability of water for additional uses
> 3. Increase the management of runoff and flows to change location, timing, and/or amount of water

Similarly, in a report by the U.S. Department of Interior (1978), *Water Conservation Opportunities Study,* the goal of water conservation is defined as ". . . the wise and judicious use of available supplies." The same ambiguity concerning the meaning of conservation prevails in a report by the ad hoc Committee on Water Resources, Commission on Natural Resources of the National Academy of Sciences. While the objective of the report was to ". . . provide guidance and assistance in formulating a water conservation research program" for the U.S. Office of Water Resources and Technology, an explicit definition of conservation was absent: in fact, their report did not distinguish between water conservation and comprehensive, efficient water supply management.

The Committee noted that water conservation consists of making better and more efficient use of water resources. But should water conservation objectives be broad enough to include better management of supplies—better hydrologic forecasts, more effective use of groundwater, more flexible facilities (such as interconnection of systems and reallocation of storage)—as well as reductions in demand? The question still stands as to what is the difference between the objectives of the water conservation program and the conventional objectives of efficient water supply planning and management.

In summary, these definitions (as has been the tradition in the history of conservation) leave something to be desired; while laudably comprehensive, they lack precision. The result is conflict at worst, confusion at best. Thus, the concept of water conservation may mean reduction of use to some, development of new supplies to others, and the curtailment of certain uses of water to yet others. To the economist, efficiency has one meaning, and to the agronomist, another. The problem of formulating a theoretically sound, yet practical definition of conservation is not new; indeed, it has plagued the conservation movement from its inception.

Since the historic Conference of Governors in 1908 in Washington, D.C., the term "conservation" has yielded to many interpretations. Gifford Pinchot, considered by many to be the father of the conservation movement in this country, stated that, "Conservation is the use of natural resources for the greatest good of the greatest number for the longest time." More specifically, Pinchot identified three objectives in conservation (Pinchot 1947):

1. Wisely to use, protect, preserve, and renew the natural resources of the earth

2. To control the use of natural resources and their products in the common interest

3. To see that the rights of people to govern themselves shall not be controlled by great monopolies through their power over natural resources

Who, indeed, would either desire or dare to disagree with such goals? However, a critical analysis of Pinchot's definition of conservation as ". . . the use of natural resources for the greatest good of the greatest number for the longest time" must conclude that, however appealing the prose, it fails as an operational definition. It will not serve as a guide in the formulation of national policy. What is the greatest good? Who determines what the greatest good is? How should it be determined? Who are the greatest number? What does "longest time" mean? How far into the future can we hope to plan?

Analysis of Pinchot's three more specific purposes of conservation also reveals difficulties. For example, the concept of conservation as wise use (as opposed, say, to one of preservation) has been interpreted in numerous ways. To some, "wise use" is interpreted to mean that renewable resources should be used before nonrenewable resources. Others feel that resources should be utilized at a rate that ensures a constant supply, as exemplified in the practice of sustained yield in forest management. Yet, other interpretations insist that the more abundant resources should be used first.

And so the arguments have run for ninety years. An agreed-upon operational definition of conservation is nonexistent. The meaning of the term *conservation* ranges from resources development to the preservation and protection of the resource base (Hays 1959). These different interpretations reflect different interests and values: to some, a resource is the physical substance itself; to others, it is its market value; and to yet others, its beauty. The question is always from what perspective the resource is being considered—economic, ecological, aesthetic, moral. Each of these worlds claim conservation as its own. Perhaps the most extreme example of conflicting stances is that provided by those who see conservation in the light (or darkness) of the Malthusian conviction that demand will outstrip resources, leading to a social catastrophe, as against those who see scarcity as the mother of research and development and thereby invention. If the former fear depletion of natural resources for future generations, the latter attempt to be reassuring by citing the empirical case of new discoveries, new technology, new skills, and thus new resources.

In the context of this history of vague, conflicting, and tendentious meanings, attempting a definition of conservation is an act of either bravery or folly; it is sailing a hazardous course through the depths of economic theory and ideological commitments. But a definition is a necessity if the management of water resources is to be informed. Hopefully, then, it is courage that accompanies this brief review of some of the more major errors to be avoided in reaching for conceptual clarity of the term conservation.

The first and greatest temptation is to define conservation from the perspective of a *single* resource. To do so is enviably but deceptively simple, for the conserving of one resource necessitates the depleting of another. It is foolishness to ignore economic relationships, for the conserving of one resource plainly necessitates the depleting of another. It is foolishness to ignore economic relationships, for none can dismiss or ignore the reality of costs. As Gordon (1958, p. 115) states:

> It seems quite plausible to suggest that if a resource is not yielding its maximum, it is being wasted. The error in the proposition can be discovered if one asks what other resources must be expended in order to achieve this maximum. . . . We cannot maximize the total product of all resources taken independently, for the respective maxima will prove to be incompatible with one another.

Still following Gordon's argument, only those attempts to provide for the future deserve to be called conservation that are not "carried out at the expense of some other form of investment." To be truly conserving, a program must demonstrate that the saving of a resource is "more productive of future incomes than alternative forms of wealth creation."

Of course, to face this fundamental economic fact takes fortitude; it requires acknowledging the awesome difficulties in identifying what the alternatives are. Moreover, it requires a tolerance for enduring ambiguity in attempting to estimate their value.

In our judgment, a second perspective on conservation to be avoided is the popular one which equates it with the "wise use of resources." Although theoretically defensible in that it assimilates the concept of conservation into the general economic problem of maximizing output, it does so at the expense of violating the everyday meaning of the word *conserve*, and prohibits distinguishing those wise uses of resources which save them for future use from those wise uses of resources which don't.

Granted, the economic interrelatedness of resources just posited argues that such a distinction is ultimately idle; that is, the *depletion* of a given resource, if wisely done, will, by definition, be an act of conservation *in the context of all resources.* Nevertheless, it is practically useful to maintain at a less-abstract level of analysis the distinction between actions that save or spend a given resource. Thus, while the term conservation might very properly be applied to a comprehensive energy program that saved gas and oil by the greater use of coal, it would be absurd in such a situation to assert that one was conserving coal.

The next pitfall to be avoided in moving toward a definition of conservation is implicit in the earlier exhortation to face the fact that conservation has costs. There we spoke of the fortitude necessary for the enormous task of determining values (costs) in order to be assured that any single act was indeed conservation. Now we must speak of another virtue—*courage,* the best word to characterize the strength needed to enter the world of values. But enter it we must. It is nonsense to insist that conservation can avoid the arena of value competition. Benefit-cost analysis is a necessary part of a definition of conservation, and values are unavoidably part of estimating benefits and costs. Rather than beating a hasty retreat, one must turn and face the field.

To do so requires analysis and assessment of all that is most intangible—aesthetics, politics, and philosophy. It requires conscious consideration of the governing values, attitudes, and beliefs of society; indeed, it requires confrontation with the sacred.

Too often both so-called conservationists and anticonservationists refuse to become involved in such insubstantial but real issues. Forty years ago, Galbraith (1958) pointed his finger at this weakness: "If we are concerned about our great appetite for materials, it is plausible to seek to increase the supply, to decrease waste, to make better use of the stocks that are available, and to develop substitutes. But what of the appetite itself? Surely this is the ultimate source of the problem. If it continues its geometric course, will it not one day have to be restrained? Yet in the literature of the resource problem this is the forbidden question." He then goes on to show that the Twentieth Century Fund, in its attempts to balance resources and use, took "present levels of consumption and prospective increases as wholly given." And that the President's Materials Policy Commission "began by stating its conviction that economic growth was important and, in degree, sacrosanct."

Galbraith is making two points here. One, of course, is that ideological denial and blindness interferes with the raising of the most pertinent conservation issues. But second, even when philosophical configurations do appear, they do so unexamined. They are "given" or "assumed" rather than deliberately identified and weighted.

But to deny or ignore or avoid ideological issues is not to disable them; they remain with their power intact. It is far more effective to formally admit them as necessary elements to any conservation decision. There are or can be real conflicts between present and future, between individuals and communities, between federal and state governments, between thrift and prodigality in consumption, between private enterprise and government control, and between the general welfare and individual freedom. And each of the antagonists in these struggles can marshal its rationale, its ethic, and its power. They are, then, factors that necessarily figure in conservation.

This argument is certainly not meant as an admonition to decision makers merely to be politic. Rather, it is a serious insistence that values must figure into the benefit-cost analysis of a conservation decision. Further, it is to emphasize—indeed, stand in awe of—the stature of the value consequences that can be involved. It is one thing to argue the merits of maintaining a canyon in its natural state versus flooding it by way of building a dam to better use water; it is quite another to debate decisions that would affect the ratio of public versus private management of resources, or lead to redistribution of wealth, or change the lifestyle of a nation. If it can reasonably be argued that different political systems have different advantages and disadvantages in times of war, it can also reasonably be argued that they have different advantages and disadvantages regarding the management of natural resources. The implication is clear: resource decisions have implications for political life. In estimating ultimate costs, it may prove frugal to place ideology above efficiency.

To be so immersed in the world of values when attempting to estimate the costs and benefits of a conservation decision is undoubtedly discomforting to those whose responsibility it is to manage the nation's resources. But an effort to escape from ideology would be as effective as a man attempting to extricate himself from quicksand by pulling on his beard. It simply can't be done. It follows, then, that any definition of conservation which holds the promise of being realistically useful must incorporate the assessment of values.

To be helpful, a definition of water conservation must possess two major attributes: it must be precise and it must be practical. A *precise* definition will permit clear distinctions to be drawn between those practices which are conservation and those which are not; a *practical* definition will facilitate the testing of specific proposed practices, so as to clearly determine whether they do indeed constitute conservation.

Since water is but one of the scarce resources required to provide water supply to users, and reductions in water use may be accompanied by increased use of other resources, not all practices that reduce use of water should be considered desirable. Only those which reduce the use, or loss,

ning assumes a very high level of system reliability, avoidance of risk, and utility ownership of all resources and facilities. The final plan often recommends construction of a large-scale water supply project. Because such planning is conducted internally, the public-at-large, outside experts, and government regulators have little or no involvement in the planning process. Consequently, the resulting plan tends to be narrowly focused and exclusionary.

Since the late 1970s, traditional supply planning has proven to be an ineffective process for some utilities. It is particularly unresponsive to the environmental, financial, and political constraints of the 1980s and 1990s; the large construction projects which are often recommended have encountered significant public opposition. It also ignores the important contribution to supply that can be obtained from comprehensive demand-side management programs such as water conservation.

*Least-cost planning* emerged during the 1980s as a response to public utility regulators who wanted utilities to provide service at the minimum cost to the ratepayer. A least-cost plan balances three interests in evaluating options: reliability, profitability, and affordability. However, in addition to analyzing typical supply alternatives for the "least cost," this form of planning also includes consideration of demand-side management alternatives in the search for minimum-cost solutions. It recognizes that demand can actually be manipulated and that the forecasted demand does not have to be taken as a given in the planning process.

In its conceptual form, the least-cost planning approach offers little specific guidance to water planners. It does not establish what costs and whose costs are to be minimized. It also implies that a single optimal solution exists and can be found despite the uncertainties of future demand growth, construction costs, environmental regulations, and other factors.

During the mid-1980s, the electricity and natural gas utilities began using a comprehensive planning approach referred to as *integrated resource planning* (IRP). Recently, water utilities took some interest in this concept in order to integrate the demand-side and supply-side alternatives. The IRP approach integrates critical planning criteria and activities into one systematic planning process. It emphasizes the least-cost principle of selecting alternatives in an attempt to minimize costs while creating a flexible plan allowing for a changing economic environment. It involves the concurrent consideration of supply and demand options including both long-term and short-term alternatives. It is conducted using an open and participatory planning process, and it emphasizes the cooperation of the many institutions involved in water resource policy and planning. It also identifies and quantifies the external costs and benefits of an alternative (known as "externalities"), and incorporates careful consideration of the uncertainties inherent in each of the alternatives.

IRP is not a well-defined procedure for developing optimal water management plans. Instead, it is a conceptual approach that must be developed further into a planning process specific to the problem and the context at hand. The key elements of IRP call for integrating planning activities which are both internal and external to the water agency. This is a key concept. In a world of scarce resources and competing policy objectives, considering the external institutional policies simultaneously with the agency's internal goals not only makes practical sense but is also politically necessary.

The basic premise of IRP is that a process is needed to integrate a wide range of traditional and innovative supply-side and demand-side alternatives. While there may be a specific alternative which addresses one objective or constraint, there will not be a single alternative which fully addresses all multiple objectives. During the IRP process, the underlying assumptions behind each objective are analyzed collectively in an attempt to correctly weight or rank each one. The rankings for each objective can then be applied to the various alternatives and analyzed numerically. The resulting IRP becomes a unique and individual product; no two IRPs are the same. Differences can occur in the numerical ranking of each of the underlying planning assumptions, objectives, and constraints, and also in the form of the public participation chosen for determining the ranking process. Thus, the IRP can be tailored to fit the particular context of the water agency and the region. Its strength as a planning tool is its flexibility, individuality, and adaptability.

It is this very individuality that makes IRP so effective. Water agencies throughout the world are facing supply issues that are very different from those they faced in the past. Plentiful supplies of water are no longer available anywhere just for the taking, even in "water-rich" regions. Environmental constraints, political realities, economic feasibility issues, and public desires have all changed the traditional supply-side planning scenarios. A sensitive planning method, like IRP, can provide guidance to the water utility, its regulators, and to the community-at-large as to the best option(s) for providing needed water supply, given the individual context of the agency and its region. Furthermore, involving the community in the determination of a solution helps to ensure that the solution can and will be effectively implemented. Because IRP is responsive to multiple needs and constraints, it is an ideal planning approach for solving regional water supply problems.

## Purpose of Volume

Since the late 1960s an enormous investment in research on urban water demand planning and management has been undertaken, especially in the United States. The products of this research effort have forced the water community to reassess traditional policies as well as identify the need for new techniques of analysis.

The purpose of this volume is to summarize the lessons learned during the past three decades and to provide direction to more efficient urban water planning and management. Concepts of demand for urban water are clarified, new methods of analyses are described and evaluated, pitfalls in estimation are highlighted, and future trends in research needs are identified. Finally, the practicability of these new methods and concepts are illustrated through several case studies in addition to step-by-step guidance in the analyses of urban water demand planning and management.

## Water Management Myths

This new perspective on urban water demand management requires continuous reexamination of assumptions and methods. Many traditional beliefs retain their validity and usefulness, but some may be found to lack a rational basis. Among these outmoded assumptions are some whose persistence in the face of evidence to the contrary qualifies them as urban water management *myths*. Five prominent myths are discussed next.

### Water Is a Necessity

It is often argued that water is not an economic good like other goods. Rather, it is claimed to be unique because it is so immediately necessary to life. According to this argument, residential uses of water, if not all urban uses, are fixed by the needs of people. They respond little, if at all, to economic incentives or other policy interventions. Water is a requirement; it is not subject to tastes, fashions, or desires.

If it were true that water is indeed a necessity, the scope for water demand management would be drastically reduced. Price would be judged ineffective as a management tool, since users are obliged to pay any price in order to obtain the water needed for survival. Water conservation would be restricted to seeking out and stopping water waste, where waste is defined as water that does not provide any service to man (leakage, for example). Perhaps demand management could be applied to certain uses, such and lawn and garden irrigation, which do not support human life. Otherwise, urban water management is confined to providing sufficient supply.

Like other urban water myths, this one rests on a simple misunderstanding. It is unarguably true that water is necessary to life. But urban water supply is not. It is possible to sustain life in the absence of a public, piped water supply. The several liters of water per day actually necessary to life can be collected by individuals and families from rivers, lakes, wells, and cisterns. Indeed, several billion of the world's inhabitants must do this every day. Urban water supply is essential to the quality of life we have become used to, not to life itself.

Since water contributes to the type of life that people lead, it can be expected that changes in the availability of water may change the benefits they receive. In this sense, water is no different from any other economic good. It is no more a necessity than food, clothing, or housing, all of which obey the normal laws of economics.

## The Myth of Per Capita Use

Just as the approaches to water supply planning and development have changed, the methods for forecasting future water requirements have also undergone a technical evolution. Historically, future water requirements were determined as the product of projected service area population and a projected value of per capita water use. However, such an approach offers only a limited account of the determinants of water use other than population. Furthermore, per capita water use is not an adequate measure of the types of water uses that occur in a community (single-family, multi-family, hotels, restaurants, petroleum refining). These uses are likely to vary among communities and over time.

Table 1-2 gives an example of how easily per capita water use estimates can misrepresent water use efficiency. City A may have a higher total municipal water use than City B, but this does *not* imply that City A uses water less efficiently than City B. Although commercial and industrial per capita water use is higher in City A, the per employee water use (a better indicator of use efficiency) is lower. This lower per employee use, yet higher per capita use in the nonresidential sector, reflects that a greater percent of the population is employed in City A than in City B (this may also reflect a larger number of commuters who work in City A relative to City B). Furthermore, per employee use rates are also a function of the types of water-using businesses and industries that exist in the city.

Note that residential per capita water use is lower in City A than in City B. This is because of the greater number of persons per household in City A than in City B (therefore City B has more housing units per capita). However, City B uses less water on a per household basis. This may be indicative of the housing mix (i.e., City B may have a larger proportion of multifamily housing units which tend to use less water per unit). Although City A has a lower residential per capita water use, its total per capita water use is higher than City B. This is because City A has a larger

**Table 1-2.** The Myth of Per Capita Use

|  | City A | City B |
|---|---|---|
| *Water Use (MGD)* | | |
| Residential | 32.317 (57%) | 32.001 (81%) |
| Commercial | 12.586 (22%) | 4.254 (11%) |
| Industrial | 11.504 (21%) | 3.102 (8%) |
| Total metered | 56.407 | 39.357 |
| *Socioeconomic Data* | | |
| Population | 237,627 | 201,054 |
| Housing units | 85,786 | 94,837 |
| Persons per household | 2.77 | 2.12 |
| Commercial employment | 112,772 | 28,571 |
| Industrial employment | 69,294 | 14,286 |
| Employee-to-population ratio | 77% | 21% |
| *Per Capita Rates of Water Use (GPCD)* | | |
| Residential | 136 | 159 |
| Commercial | 53 | 21 |
| Industrial | 48 | 15 |
| Total metered | 237 | 195 |
| *Other Rates of Water Use* | | |
| Residential (gallons/housing unit/day) | 377 | 337 |
| Commercial (gallons/employee/day) | 122 | 148 |
| Industrial (gallons/employee/day) | 166 | 217 |

commercial/industrial base than City B (i.e., a greater number of persons employed).

Failure to take into account major influences on future water use in various sectors, such as changes in income, housing stock, industrial mix, and the price of water, are the most critical shortcomings of the per capita method of forecasting water demand. The per capita approach can seriously overestimate demand for water, thereby resulting in unnecessary and costly investments. Over the last two decades, water planners have begun using disaggregated water use forecasts, which take into account differences in the socioeconomic characteristics of the resident population, as well as seasonal variations in economic and climatic conditions of a study area. *The IWR-MAIN Water Demand Analysis Software* was developed as a water-planning tool that provides a thorough, disaggregated approach to water-demand forecasting and analysis (see Chapter 4).

Another example that illustrates the effects of specific determinants of water use and the misrepresentation of per capita use is Table 1-3. In 1991, the Metropolitan Water District of Southern California developed disag-

gregate forecasts of water use for 57 study areas which in total, represented their entire service area (Dziegielewski and Opitz 1991). These
forecasts not only addressed the major customer groups within the service
area but also addressed the factors that are expected to affect average
water use rates in the future: These factors are summarized in Table 1-3.
Overall, water use per capita in Metropolitan's water service area is
expected to increase between 1990 and 2010 by 24.4 GPCD. Almost half
(11.3 GPCD) of this increase is the result of the projected decrease in number of persons per household and the increase of the percentage of the
population employed. Differential growth trends throughout the service
area, including the increasing rate of inland growth, account for 8.2 GPCD
of the increase in per capita rates. Projected changes in home value distributions (resulting from increasing real household income) is the next
major factor, accounting for 6.2 GPCD of the increase. Changes in housing
and employment mixes represent a 1.3 GPCD decrease over the 20-year
period. Each of these effects is discussed in greater detail below.

**Inland Growth.**   Between 1990 and 2010, the population of the Metropolitan service area was projected to increase by approximately 3.5 million. Population will increase disproportionately in desert (inland) areas
of San Bernardino and Riverside counties. Because these areas are characterized by higher air temperatures and lower precipitation in comparison
to the rest of the service area, water requirements in new inland developments are expected to be higher than average. Overall, the inland growth
is expected to increase the per capita use rate by 5.0 GPCD.

**Geographic Growth Differentials.**   In addition to the inland growth
effect, the distribution of new population among the existing urbanized
areas will not be proportional. Population projections indicate that more
new homes and economic activities will be located in urban clusters
which are characterized by higher-than-average per capita water use. This
trend will result in a 3.2 GPCD increase in the areawide water use.

**Standard of Living.**   Median household income in Southern California
counties was projected to increase during the planning period by 8 to 28
percent above the 1990 levels (in real dollars). The growth in income is
expected to enhance the standard of living as expressed by the presence of
convenience products in the house (e.g., washing machine, dishwasher,
garbage disposal, multiple bathrooms, evaporative cooler, humidifier)
and decorative or convenience outdoor features and facilities (e.g., lawn,
flower beds, decorative beds, automatic sprinkling systems). These home
improvements are expected to result in a 6.2 GPCD increase in use rates
between 1990 and 2010.

**Table 1-3.** Factors Affecting Changes in Base Per Capita Water Use (GPCD)

| Sector | Base per capita water use (GPCD) 1990 | Income 2010 | Factors affecting per capita use (changes in GPCD between 1990–2010) Housing change | Mix effect | Family (&MF/SF) | Industry size | Labor mix | Inland force | County Differential growth[a] | Differential growth[b] |
|---|---|---|---|---|---|---|---|---|---|---|
| Residential | 144.6 | 159.9 | +15.3 | +6.2 | -3.3 | 8.8 | — | — | +2.1 | +1.5 |
| Commercial | 37.4 | 42.9 | +5.5 | — | — | — | +2.3 | +2.7 | +0.4 | +0.1 |
| Industrial | 12.2 | 12.2 | 0.0 | — | — | — | -0.3 | -0.2 | +0.6 | -0.1 |
| Other | 22.9 | 26.5 | +3.6 | — | — | — | — | — | +1.9 | +1.7 |
| Total M & I | 217.1 | 241.6 | +24.4 | +6.2 | -3.3 | 8.8 | +2.0 | +2.5 | +5.0 | +3.2 |

[a] Represents growth shifting to hotter, drier inland desert areas.
[b] Caused by faster growth within county in areas with higher base water use.

25

**Housing Type Mix.**   The increasing share of multifamily housing units in the total housing stock will tend to decrease per capita water use. In 1990, about 44 percent of all housing units were other than single-family (i.e., duplexes, triplexes, apartments). By 2010, the share of multifamily housing units was projected to increase to 48 percent. Because multifamily structures share landscaping and swimming pools and generally have fewer water-using appliances (e.g., washing machines or dishwashers), the average water use is lower than in detached single-family residences. The areawide effect of this trend is expected to decrease the per capita water use by 3.3 GPCD.

**Household Size.**   There is a trend in Southern California and in the rest of the country toward decreasing household size. The result of this trend is that more housing units will be built for each thousand persons of population increase in the future than in the past. In Southern California, the ratio of total population to total number of occupied housing units was 2.85 in 1990. By 2010, this ratio was projected to decrease to 2.69. Because some of the water uses in a household remain constant regardless of the number of household members (e.g., landscaping) and because some economies of size cannot be achieved in smaller households (e.g., full loads in clothes washers and dishwashers), this trend is expected to increase the per capita water use between 1990 and 2010 by 8.8 GPCD.

**Labor Force Participation.**   A greater share of the population being employed in the next two decades will cause the gross regional per capita product to increase. The fraction of population employed was projected to increase from 48 percent in 1990 to 50 percent in 2010. The expected effect of this trend is a 2.5 GPCD increase in water use by 2010.

**Mix of Economic Activities.**   The distribution of economic activities is constantly changing, thus affecting the use of water in commercial and manufacturing sectors. National, state, and regional projections suggest that the commercial sectors with the highest per employee water use— hotels, amusements, hospitals—will provide an expanding share of future employment. In Southern California, this trend is expected to result in a 2.0 GPCD increase in water use.

It should be noted that the previous example represents a baseline water use forecast and does not take into account the impacts of water demand management programs that have been in place since 1990 or will be implemented in the service area.

## Water Users Do Not Respond to Price

Traditional approaches to urban water management rest on a number of stated and unstated assumptions. For example, many claim that water use does not respond to changes in price. Various reasons have been given for this opinion. One, discussed earlier, is the myth that water is a necessity and therefore not a normal economic good. Another theory is that water use is matter of habit, where most water users do not examine or analyze activities that use water. According to this view, changes in water price would be irrelevant to users who do not consciously associate their routine daily activities with water use. Still another idea is that users would be capable of responding to price, but do not do so for various reasons. The reasons include the small size of the water bill as a fraction of the household budget, the even smaller impact of price changes, and the difficulty of associating specific changes in water use behavior with changes in the total cost of water.

This opinion is validated in the minds of many by their personal observations of water use changes over time. Increases in water price are sometimes associated with increases, rather than reductions, in water use in the following years. Such casual observations fail to recognize that water use may increase in response to growing population and economic activity, changing weather, and other factors.

The most straightforward response to this myth is to note that water use does, in fact, respond to changes in price. When sufficient data are collected and controlled for other influences on water use, the effect of price emerges quite clearly. The record of empirical studies is unambiguous on this point. A 1984 survey (Boland et al. 1984) reviewed more than 50 studies of urban water use which considered the possibility of a price effect. While the observed sensitivity of water use to changes in price (the price elasticity of demand) varies substantially from study to study, no study concluded that price had no effect on water use. Since 1984, at least 50 additional studies have appeared in the literature, documenting price effects in still more places and other various conditions. The overall conclusion is the same: no study in the peer-reviewed literature concludes that price does not affect urban water use.

## The Myth of Excessive Irrigation of Urban Landscapes

There is a widely held belief among water professionals that in many urban areas, especially in the West, high water use is a result of overwatering of residential landscapes. In reality, the average rates of water use for irrigation purposes indicate that urban landscapes are subject to deficit irrigation. Kiefer and Dziegielewski (1991) examined irrigation rates in a sample of 515 southern California households and concluded that during summer approximately 60 percent of households applied less water than the theoretical agronometric requirement. The magnitude of the deficit was approximately 205 gallons per day, or 58 percent below the requirement. Only 40 percent of households applied enough or more than enough water assuming that their irrigation methods were 100 percent efficient in delivering water to the plants. The average for all sample households was an irrigation deficit of 40 gallons per day (or 11 percent of the theoretical requirement).

These results clearly show that urban householders do not overwater their landscapes. In the eastern part of the country, where rainfall is generally sufficient to maintain grass and ornamental plants, the irrigation is incidental and on average accounts for only 20 to 30 gallons per day of water use during summer (Davis, Beezhold, Opitz, and Dziegielewski 1996).

## Conservation Will Lead to Negative Fiscal Impacts on the Water Supply Agency

Water supply agencies are expected to serve social objectives related to the rational management of resources; at the same time, they must remain viable as business enterprises. A policy of promoting the conservation of water strikes many as a conflict between these roles. Water conservation may be beneficial from a resource management perspective, but the supply agency fears revenue losses and earnings shortfalls. It is often assumed that these fiscal problems can be cured only by substantial increases in user charges, with resulting consumer opposition and political interference. There is a particular sensitivity to the possibility of a rate increase, which appears to punish consumers for their cooperation in reducing water use.

While there have been instances of significant rate increases following conservation initiatives, with all the negative reaction that one might expect, this is by no means a necessary outcome. As discussed previously, water conservation measures are intended to provide a beneficial reduction in water use. The benefits of lower water use levels are expected to exceed the costs. In most cases, these benefits derive from lower operating costs and deferred facility costs. From the water supply agency's point of view, conservation means lower costs (benefits) and reduced revenue (costs). Over time, the benefits should exceed the costs, meaning that the agency will be in a better financial position as a result of conservation. If that is not true, one might question the design of the conservation program.

Still, there is a need for careful financial planning. The largest benefits are generally those associated with deferred costs of future facilities. These cost savings will not appear immediately; in fact, the costs that are to be saved may not yet be incorporated into current water charges. When conservation measures are imposed during a drought, the utility may have other drought response costs that are not reflected in rates. Also, the examination and reflection that accompany the decision to promote conservation may uncover other areas in which the current charges are deficient. In other words, the current charges may be too low, with or without conservation. If conservation is implemented first and the charges are then adjusted, consumers will assume that it was their conservation efforts that caused the higher charges. The cure for this, of course, is to ensure that user charges are set at the correct level prior to implementing conservation.

Finally, from the consumer's point of view, it should be noted that the more or less simultaneous appearance of water conservation measures and higher user charges does not necessarily imply an increased cost burden. Only customers who fail to respond to the conservation incentives will see increased bills. The typical customer will pay a higher unit charge for a smaller quantity of water, but the total bill will also be smaller (reflecting the lower costs incurred by the utility).

## References

American Water Works Association (AWWA). (undated) "Operating Data for Water Utilities 1970 and 1965." AWWA statistical report no. 20112. New York.

American Water Works Association (AWWA). 1981. "1981 Water Utility Operating Data."

American Water Works Association (AWWA). 1992. "Water Industry Data Base: Utility Profiles."

Babbit, Harold E. and James J. Doland. 1929. *Water Supply Engineering.* New York: McGraw-Hill.

Baker, M. N. 1948. *The Quest for Pure Water.* New York: American Water Works Association.

Baumann, D. D., J. J. Boland, J. H. Sims, B. Kranzer, and P. H. Carver. *The Role of Conservation in Water Supply Planning.* Fort Belvoir, VA: U.S. Army Corps of Engineers, Institute for Water Resources, in press.

Boland, John J., Benedykt Dziegielewski, Duane D. Baumann, and Eva M. Opitz. 1984. "Influence of Price and Rate Structures on Municipal and Industrial Water Use," IWR Report 84-C-2. Fort Belvoir, VA: U.S. Army Corps of Engineers, Institute for Water Resources.

Commission on Natural Resources, ad hoc Committee on Water Resources. Water Conservation Research. A Report to the Office of Water Research and Technology. Washington, D.C.: National Academy of Sciences, October 2, 1978 (mimeo).

Davis, W. Y., M. T. Beezhold, E. M. Opitz, and B. Dziegielewski. 1996. ACT-ACF Comprehensive Study Municipal and Industrial Water Use Forecasts. volume I: Technical Report. PMCL. Carbondale, IL.

Dziegielewski, B. and E. Opitz. 1991. *Municipal and Industrial Water Use in the Metropolitan Water District Service Area: Interim Report No. 4.* Carbondale, IL: Planning and Management Consultants, Ltd.

Frontinus, Sextus Julius. 1899. *The Two Books of the Water Supply of the City of Rome,* transl. by Clemens Herschel. Boston: Dana Estes and Co.

Galbraith, J. "How Much Should A Country Consume?" 1958. In *Perspectives On Conservation.* Henry Jarrett, ed. Baltimore: The Johns Hopkins Press.

Gordon, Scott. "Economics and the Conservation Questions." *The Journal of Law and Economics* 1 (October 1958): 110–121.

Hays, Samuel P. *Conservation and the Gospel of Efficiency.* 1959. Cambridge: Harvard University Press.

Keiffer, J. C. and B. Dziegielewski. 1991. Analysis of Residential Landscape Irrigation in Southern California. Research report prepared for Metropolitan Water District of Southern California, Los Angeles, CA, December.

National Academy of Sciences. *Potential Technological Advances and Their Impact on Anticipated Water Requirements.* Washington: National Academy of Sciences. June 1971.

Pinchot, Gifford. 1947. *Breaking New Ground.* New York: Harcourt, Brace and Co.

Sewerage Commission of the City of Baltimore. 1897. *Report of the Sewerage Commission of the City of Baltimore.* Baltimore: The Friedenwald Company.

United States Department of the Interior. *Report on the Water Conservation Opportunities Study.* Bureau of Reclamation and Bureau of Indian Affairs. September 1978.

U.S. Water Resources Council. *The Nation's Water Resources, The Second National Assessment.* Washington, D.C.: U.S. Water Resources Council. October 1977.

U.S. Bureau of the Census. 1995. *Annual Capital Expenditures: 1993.*

U.S. Bureau of the Census. 1996a. *1992 Census of Manufacturers.*

U.S. Bureau of the Census. 1996b. *1992 Census of Mineral Industries.*

U.S. Bureau of the Census. 1996c. *1992 Census of Transportation, Communications, and Utilities, U.S. Summary.*

U.S. Bureau of the Census. 1996d. *Annual Capital Expenditures: 1994.*

U.S. Bureau of the Census. 1996e. *Annual Survey of Government Finances: FY 93.*

U.S. Department of Commerce. 1996. "National Income and Product Accounts Tables," *Survey of Current Business,* August, pp. 15–80.

Vogan, Christine R. 1996. "Pollution Abatement and Control Expenditures, 1972–94," *Survey of Current Business,* September, pp. 48–67.

Washington Suburban Sanitary Commission, *A Customer Handbook on Water-Saving and Wastewater-Reduction,* 1972.

In this case $\alpha_i$ and $\beta_i$ are treated as constants—they vary by industry $i$ but are fixed over all establishments in an industry. An example would be to use the preceding data on average intake per employee in various industries, with $\beta_i = 9392$ for petroleum refining and $\beta_i = 3454$ for the paper industry, and so forth. A more sophisticated version relaxes the assumption of strict proportionality and postulates

$$x_i = \alpha_i y_i^{\gamma} \tag{3}$$

$$x_i = \beta_i E_i^{\gamma} \tag{4}$$

where $\gamma$ may or may not vary with industry $i$. Water use increases less than proportionately with scale of production if $\gamma < 1$, and more than proportionately if $\gamma > 1$. For U.S. manufacturing industry generally, Dziegielewski (1988) found that a value of $\gamma = 0.7$ fit the data well. It should be noted that, when $\gamma \neq 1$, the $\alpha_i$'s and $\beta_i$'s in Equations (3) and (4) will take different numerical values than the $\alpha_i$'s and $\beta_i$'s in Equations (1) and (2).

Economists criticize this approach on two grounds. The first criticism is that it treats the parameters $\alpha$, $\beta$, and $\gamma$ as constant across all firms within a given industry regardless of other circumstances. It ignores other factors that can affect water use within an industry, including economic variables such as the cost of water relative to that of other inputs which might be complements or substitutes for water in the production process. The second criticism is that this approach treats the right-hand-side variable— output ($y$) or employment ($E$)—as exogenous to the firm's decision on water use. It assumes that the firm takes as given its level of output or the size of its labor force when it determines how much water to use. This seems implausible: firms are more likely to determine their level of output or the size of their labor force in concert with their level of water use. In that case, the analyst should model the simultaneous determination of both $x$ and $E$ or $y$, rather than modeling $x$ taking the other variables as given.

Economic theory provides a way to do this by embedding a firm's demand for inputs in a formal model of optimization by the firm. The optimization is usually broken into two stages. The first stage is how to produce a given target output in the best manner. The second stage is what scale of output to target. At the first stage, the level of output is taken as given, but all inputs, including water use and the size of the labor force, are determined jointly. At the second stage, the level of output is selected together with the corresponding inputs. The result is that there are two types of demand function for water in economic theory, one corresponding to the first stage, called the *fixed-output* or *conditional demand function,* and the other corresponding to the second stage, called the *variable-output* or *unconditional demand function* for water. The demand-forecasting Equations (2) and (4) are not consistent with either type of demand function; Equations (1) and (3) are consistent with the first type, the fixed-output demand function, but not the second.

There can be circumstances in which the first type of demand function is likely to be more realistic in that the firm is not really free to determine its own output. For example, investor-owned public utilities regulated by state Public Utility Commissions (PUC) are often required to meet all the demand occurring within their service area at prices set by the PUC.* In those cases, output really is exogenous for the firm; it can't set its own output either directly, by selecting a particular level of production, or indirectly, by choosing prices calculated to induce a particular level of demand. Under those circumstances it makes sense to model the demand for inputs taking output as given, yielding conditional demand functions for water and other inputs. In other circumstances, however, it is likely that the firm does determine its own level of production, so that the unconditional demand functions are more relevant for empirical analysis.

Another basic distinction in economic theory is that between short-run and long-run input demand functions. The notion is that it can take more time to vary some inputs than others; plant and capital equipment, for example, may take longer to change than inputs such as water, energy, materials, or the size of the labor force. Therefore, for a certain span of time, plant and capital equipment are fixed inputs while the others are variable inputs. The *short run* is defined as the period of time during which some inputs are fixed; the *long run* is defined as when all inputs are variable. Each of the two types of demand function exists in both the short and long run. What is described earlier is the long-run version, in which all inputs are assumed to be variable. The short-run version of the two stages of firm decision making is that, in the first stage, the firm decides how best to produce a given target output taking the fixed inputs as given but freely varying the variable ones; and, in the second stage, it selects the optimal level of output together with the corresponding variable inputs, taking the fixed inputs as given. Thus, there are short- and long-run conditional and unconditional demand functions.

To illustrate the differences among the four types of demand function, and demonstrate the connections among them, it is helpful to introduce some formal notation. Consider an establishment producing an output through the use of certain inputs. Denote the volume of output produced per unit of time by $y$. Suppose that the firm uses $N$ inputs, one of which is water. Let $x_k$ denote the amount of the $k$th input used per unit of time,

* This would include investor-owned water utilities.

$k = 1, \ldots, N$. If the first input is water, then $x_1$ is the amount of water used by the firm. Let $w_k$ be the price of the $k$th input, $k = 1, \ldots, N$, and let $p$ be the price at which the firm sells its output. Then the firm's total cost of production is given by

$$C = w_1 x_1 + \cdots + w_N x_N \equiv \sum_{k=1}^{N} w_k x_k \tag{5}$$

and its total profit from production is

$$\pi = py - \sum_{k=1}^{N} w_k x_k \tag{6}$$

We assume that the production technology available to the firm can be represented by some function relating inputs to output

$$y = f(x_1, \ldots, x_N) \tag{7}$$

which is known as the production function. For each possible combination of inputs, the production function states the corresponding quantity of output that will be obtained. This information may be conveyed in the form of an algebraic equation, a graph, or a table. For example, one particular form known as the Cobb-Douglas production function is represented by the equation

$$y = A x_1^{\alpha_1} x_2^{\alpha_2} \cdots x_N^{\alpha_N} \tag{8}$$

where $A > 0$ and $\alpha_k > 0$, $k = 1, \ldots, N$, are coefficients whose numerical values would be estimated statistically from data. Another formula is the constant elasticity of substitution (CES) production function, given by

$$y = A^{\mu} (\delta_1 x_1^{-\rho} + \delta_2 x_2^{-\rho} + \cdots + \delta_N x_N^{-\rho})^{-\mu/\rho} \tag{9}$$

where $A > 0$, $\mu > 0$, $\rho > -1$, and $\delta_k > 0$, $k = 1, \ldots, N$, are coefficients that would be estimated from data, with $\sum_k \delta_k = 1$.

The production function is a representation of technology rather than behavior. The firm's behavior can be characterized in different ways, depending on what choices it has and what are its objectives for those choices. It is natural to focus on the two choices mentioned previously, corresponding to two levels of firm decision making. The first decision is cost minimization: how to produce a given output target at minimum cost. For the moment, we consider the long-run case where all $N$ inputs are variable. In this case, the firm's problem is to select $x_1, \ldots, x_N$ so as to

$$\text{minimize} \quad C = \sum_{k=1}^{N} w_k \cdot x_k \qquad \text{subject to } y = f(x_1 \cdots x_N) \tag{10}$$

The solutions for $x_1, \ldots, x_N$ are the optimal input levels the firm should use when it is faced with input prices $w_1, \ldots, w_N$ and aims to produce an output of $y$ at minimum cost. The behavioral rule relating the givens in the problem—$w_1, \ldots, w_N$ and $y$—to the optimal choice of inputs is a set of functions of the form

$$x_k = g^k(w_1, \ldots, w_N, y) \qquad k = 1, \ldots, N \tag{11}$$

These are the long-run conditional input demand functions. The resulting cost of production for the firm at the optimum is

$$C = \sum_{k=1}^{N} w_k g^k(w_1 \cdots w_N, y) \equiv C(w_1 \cdots w_N, y) \tag{12}$$

which is the firm's total cost function. For example, with the Cobb-Douglas production function (8) when $N = 2$ ($x_1$ is the input of water, say, and $x_2$ represents all other inputs combined), the conditional input demand functions take the form*

$$x_1 = g^1(w_1, w_2, y) = \left(\frac{y}{A}\right)^{1/(\alpha_1 + \alpha_2)} \left(\frac{\alpha_1 w_2}{\alpha_2 w_1}\right)^{\alpha_2/(\alpha_1 + \alpha_2)}$$

$$x_2 = g^2(w_1, w_2, y) = \left(\frac{y}{A}\right)^{1/(\alpha_1 + \alpha_2)} \left(\frac{\alpha_2 w_1}{\alpha_1 w_2}\right)^{\alpha_1/(\alpha_1 + \alpha_2)} \tag{13a}$$

and the cost function is

$$C = c(w_1, w_2, y) = (\alpha_1 + \alpha_2) \left(\frac{y}{A}\right)^{1/(\alpha_1 + \alpha_2)} \left(\frac{w_1}{\alpha_1}\right)^{\alpha_1/(\alpha_1 + \alpha_2)} \left(\frac{w_2}{\alpha_2}\right)^{\alpha_2/(\alpha_1 + \alpha_2)} \tag{13b}$$

When $N = 2$ with the CES production function (9) the conditional input demand functions take the form

$$x_1 = \frac{y^{1/\mu}}{A} \delta_1^{1/\rho} \left[\left(\frac{\delta_2}{\delta_1}\right)^{\sigma} \left(\frac{w_2}{w_1}\right)^{\sigma\rho} + 1\right]^{1/\rho}$$

$$x_2 = \frac{y^{1/\mu}}{A} \delta_2^{1/\rho} \left[\left(\frac{\delta_1}{\delta_2}\right)^{\sigma} \left(\frac{w_1}{w_2}\right)^{\sigma\rho} + 1\right]^{1/\rho} \tag{14a}$$

and the cost function is

$$C = \frac{y^{1/\mu}}{A} [\delta_1^{\sigma} w_1^{\sigma\rho} + \delta_2^{\sigma} w_2^{\sigma\rho}]^{1/\sigma\rho} \tag{14b}$$

---

* More detailed discussion and proofs of results stated in this chapter can be found in intermediate or advanced microeconomic theory text books, such as Varian (1992 or 1996).

where $\sigma \equiv 1/(1 + \rho) > 0$ is what is known as the elasticity of substitution parameter.

The derivatives of the demand functions in (11) describe the sensitivity of the conditional demand for an input $x_i$, to a change in its own price $w_i$, the prices of other inputs $w_j (j \neq i)$, or the scale of production $y$. When expressed in percentage terms, these derivatives are known as *elasticities:* an elasticity measures the proportional change in one variable (in this case input demand) per 1 percent change in another variable (e.g., input price). For conditional demand functions, the own-price elasticity of demand is defined as

$$_c\epsilon_i^i \equiv \frac{w_i}{x_i} \frac{\partial g^i(w_1, \ldots, w_N, y)}{\partial w_i}$$

the cross-price elasticity of demand is defined as

$$_c\epsilon_j^i = \frac{w_j}{x_i} \frac{\partial g^i(w_1, \ldots, w_N, y)}{\partial w_j}$$

and the output elasticity of demand is defined as

$$_c\epsilon_y^i \equiv \frac{y}{x_i} \frac{\partial g^i(w_1, \ldots, w_N, y)}{\partial y}$$

It can be shown that the own-price derivative of a conditional demand function must be nonpositive—that is, it can be negative or zero but not positive. If the own-price elasticity is −0.3, for example, this means that a 10 percent increase in the price of an input results in a 3 percent reduction in the amount of it demanded, holding other input prices and the scale of output constant. If an own-price elasticity is less than unity in absolute value, the demand for the input is said to be inelastic;* if the elasticity is greater than unity in absolute value, the demand is said to be elastic. In contrast, the cross-price derivatives, and the cross-price elasticities, can be negative or positive; if the price of the $j$th input increases by 10 percent, the demand for the $i$th input may either decrease or increase. If the demand for the $j$th input decreases (i.e., the cross-price derivative is negative), inputs $i$ and $j$ are said to be complements; if demand for the $j$th input increases (i.e., the cross-price derivative is positive), inputs $i$ and $j$ are said to be substitutes. In a manufacturing industry, for example, water

---

* It follows by definition that, if the elasticity of demand for an input is less than unity, total expenditure on the input rises when the price of the input increases, and falls when the price decreases. If the elasticity of demand exceeds unity, the opposite is true: total expenditure on the input falls when the price rises and rises when the price falls.

and labor may be substitutes but water and capital may be complements;
if so, the use of water rises when labor becomes more expensive but falls
when capital becomes more expensive, holding the target output constant.
The output derivative can also have either sign. If it is positive, so that the
amount of input demanded increases when the target output increases,
the input is said to be normal; if it is negative, so that input demand falls
when the target output increases, the input is inferior. It is impossible for
all inputs to be inferior: you have to use more of at least some inputs to
produce more output. But, it is possible that some inputs can be inferior—
it is cost-effective for the firm to use them at low scales of production but,
as the scale of production rises, the firm substitutes away from them to
other inputs such that their use falls in absolute as well as relative terms.
However, water is generally unlikely to be an inferior input in the manufacturing industry.

With the Cobb-Douglas input demand functions in (13a), the demand
elasticities are

$$_c\epsilon_i^i = -\alpha_2/(\alpha_1 + \alpha_2)$$

$$_c\epsilon_j^i = \alpha_2/(\alpha_1 + \alpha_2)$$

$$_c\epsilon_y^i = 1/(\alpha_1 + \alpha_2)$$

In this case, the two inputs are substitutes and each input is normal. It is
a general feature of the Cobb-Douglas production function that it requires
all inputs to be substitutes. By contrast, the CES production function
allows inputs to be complements or substitutes. Like the Cobb-Douglas,
the CES production function requires all inputs to be normal. However,
there are other production functions, such as the translog discussed below,
which permit inferior inputs.

The other decision by the firm is profit maximization: how much of each
input to use and how much output to produce for maximum profit. Again,
we first assume long-run profit maximization, where all $N$ inputs are variable. In this case, the firm's problem is to select $x_1, \ldots, x_N$ so as to

$$\text{maximize} \quad \pi \equiv pf(x_1 \cdots x_N) - \sum w_k x_k \tag{15}$$

The solutions for $x_1, \ldots, x_N$ are the optimal input levels the firm should
use to maximize profit when faced with input prices $w_1, \ldots, w_N$ and output price $p$. The behavioral rule relating the givens in the problem to the
optimal choice of inputs is a set of functions of the form

$$x_k = h^k(w_1 \cdots w_N, p) \qquad k = 1 \cdots N \tag{16}$$

These are the long-run unconditional input demand functions. The optimal level of output which the firm should supply is given by

$$y = f[h^1(w_1 \cdots w_N, p), \ldots, h^N(w_1 \cdots w_N, p)] \equiv y(w_1 \cdots w_N, p) \quad (17)$$

which is the firm's long-run output supply function. The firm's profit at the optimum is

$$\pi = p \cdot y(w_1 \cdots w_N, p) - \sum w_k h^k(w_1 \cdots w_N, p) \equiv \pi(w_1 \cdots w_N, p) \quad (18)$$

which is known as the firm's total profit function. For example, with the Cobb-Douglas production function (8), the unconditional input demand functions take the form

$$x_1 = h^1(w_1, w_2, p) = A^{1/(1-\alpha_1-\alpha_2)} p^{1/(1-\alpha_1-\alpha_2)} \left(\frac{w_1}{\alpha_1}\right)^{(\alpha_2-1)/(1-\alpha_1-\alpha_2)} \left(\frac{w_2}{\alpha_2}\right)^{-\alpha_2/(1-\alpha_1-\alpha_2)}$$

$$x_2 = h^2(w_1, w_2, p) = A^{1/(1-\alpha_1-\alpha_2)} p^{1/(1-\alpha_1-\alpha_2)} \left(\frac{w_1}{\alpha_1}\right)^{-\alpha_1/(1-\alpha_1-\alpha_2)} \left(\frac{w_2}{\alpha_2}\right)^{(\alpha_1-1)/(1-\alpha_1-\alpha_2)} \quad (19a)$$

the output supply function is

$$y = y(w_1, w_2, p) = A^{1/(1-\alpha_1-\alpha_2)} p^{\alpha_1+\alpha_2/(1-\alpha_1-\alpha_2)} \left(\frac{w_1}{\alpha_1}\right)^{-\alpha_1/(1-\alpha_1-\alpha_2)} \left(\frac{w_2}{\alpha_2}\right)^{-\alpha_2/(1-\alpha_1-\alpha_2)} \quad (19b)$$

and the profit function is

$$\pi = \pi(w_1, w_2, p)$$
$$= (1-\alpha)[A\alpha_1^{\alpha_1}\alpha_2^{\alpha_2}]^{1/(1-\alpha_1-\alpha_2)} p^{1/(1-\alpha_1-\alpha_2)} w_1^{-\alpha_1/(1-\alpha_1-\alpha_2)} w_2^{-\alpha_2/(1-\alpha_1-\alpha_2)} \quad (19c)$$

With the CES production function, (9), the unconditional input demand functions take the form

$$x_1 = A^{\mu/(1-\mu)} \mu^{1/(1-\mu)} p^{1/(1-\mu)} \delta_1^{\sigma} w_1^{-\sigma} [\delta_1^{\sigma} w_1^{\sigma\rho} + \delta_2^{\sigma} w_2^{\sigma\rho}]^{-(\rho+\mu)/[\rho(1-\mu)]}$$

$$x_2 = A^{\mu/(1-\mu)} \mu^{1/(1-\mu)} p^{1/(1-\mu)} \delta_2^{\sigma} w_2^{-\sigma} [\delta_1^{\sigma} w_1^{\sigma\rho} + \delta_2^{\sigma} w_2^{\sigma\rho}]^{-(\rho+\mu)/[\rho(1-\mu)]} \quad (20a)$$

the output supply function is

$$y = (A\mu)^{\mu/(1-\mu)} p^{1/(1-\mu)} [\delta_1^{\sigma} w_1^{\sigma\rho} + \delta_2^{\sigma} w_2^{\sigma\rho}]^{-\mu/[(1-\mu)\sigma\rho]} \quad (20b)$$

and the profit function is

$$\pi = (1-\mu)(A\mu)^{\mu/(1-\mu)} p^{1/(1-\mu)} [\delta_1^{\sigma} w_1^{\sigma\rho} + \delta_2^{\sigma} w_2^{\sigma\rho}]^{-\mu/[(1-\mu)\sigma\rho]} \quad (20c)$$

The derivatives of (16) describe the sensitivity of the unconditional demand for an input $x_i$ to changes in its own price $w_i$, changes in the prices of other inputs $w_j (j \neq i)$, or changes in the price of output $p$. As with the derivatives of the conditional demand function, these can be expressed in percentage terms as elasticities. For unconditional demand functions, the own-price elasticity of demand is defined as

$$\epsilon_i^i \equiv \frac{w_i}{x_i} \frac{\partial h^i(w_1, \ldots, w_N, p)}{\partial w_i}$$

the cross-price elasticity of demand is defined as

$$\epsilon_j^i \equiv \frac{w_j}{x_i} \frac{\partial h^i(w_1, \ldots, w_N, p)}{\partial w_j}$$

and the output price elasticity of demand is defined as

$$\epsilon_p^i \equiv \frac{p}{x_i} \frac{\partial h^i(w_1, \ldots, w_N, p)}{\partial p}$$

In addition, the derivatives of (17) describe the sensitivity of output supplied to changes in output price $p$, or the price of an input $w_i$. These, too, can be expressed in percentage terms as elasticities. The output price elasticity of supply is defined as

$$\epsilon_p^y \equiv \frac{p}{y} \frac{\partial y(w_1, \ldots, w_N, p)}{\partial p}$$

and the input price elasticity of supply is defined as

$$\epsilon_i^y \equiv \frac{w_i}{y} \frac{\partial y(w_1, \ldots, w_N, p)}{\partial w_i}$$

It can be shown that the output price elasticity of supply is generally positive—an increase in output price triggers some increase in the amount of output supplied—but the input price elasticity of supply can be of either sign. Specifically, it turns out that there is a mathematical relationship between whether an input is normal or inferior and how a change in its price affects the output supply curve. The result of this relationship is that, if an input is normal, an increase in its price reduces the amount of output supplied; if it is inferior, an increase in its price raises the amount of output supplied. In terms of elasticities,

$$_c\epsilon_y^i \cdot \epsilon_i^y < 0 \tag{21}$$

So far we have focused on long-run demand functions. To define the corresponding short-run versions of these functions, assume that the

$N$th input is something which is fixed in the short run, such as capital equipment, while inputs $1, \ldots, N - 1$ are variable in the short run. Suppose that capital is fixed at a level of $\bar{x}_N$. Then, the total fixed costs of production are TFC $= w_N \cdot \bar{x}_N$, and the total variable costs of production are TVC $= \sum_{k=1}^{N-1} w_k \cdot x_k$. Total cost is the sum of total fixed cost and total variable cost. Short-run cost minimization involves selecting variable inputs $x_1, \ldots, x_{N-1}$ so as to minimize total variable cost

$$\text{minimize} \quad C = \sum_{k=1}^{N-1} w_k \cdot x_k \quad \text{subject to } y = f(x_1 \cdots x_{N-1}, \bar{x}_N) \quad (22)$$

The solution for $x_1, \ldots, x_{N-1}$ is the optimal levels of the variable inputs for producing a fixed output $y$, given the levels of the fixed inputs. The short-run conditional demand functions take the form

$$x_k = g^k(w_1, \ldots, w_{N-1}, \bar{x}_N, y) \quad k = 1, \ldots, N - 1 \quad (23)$$

Given cost minimization, the short-run total variable cost function is

$$C = \sum w_k g^k(w_1, \ldots, w_{N-1}, \bar{x}_N, y) \equiv C(w_1 \cdots w_{N-1}, \bar{x}_N, y) \quad (24)$$

Similarly, short-run profit maximization involves selecting variable inputs $x_1, \ldots, x_{N-1}$ so as to

$$\text{maximize} \quad \pi \equiv f(x_1 \cdots x_{N-1}, \bar{x}_N) - \sum_{k=1}^{N-1} w_k \cdot x_k - w_N \cdot \bar{x}_N \quad (25)$$

The solution is the optimal levels of the variable inputs for maximizing profits given input and output prices $w$ and $p$, and given the levels of the fixed inputs. The short-run unconditional demand functions take the form

$$x_k = h^k(w_1 \cdots w_{N-1}, \bar{x}_N, p) \quad k = 1 \cdots N - 1 \quad (26)$$

The optimal output supply in the short run is

$$y = f[h^1(w_1 \cdots w_{N-1}, \bar{x}_N, p), \ldots, h^{N-1}(w_1 \cdots w_{N-1}, \bar{x}_N, p), \bar{x}_N]$$

$$\equiv y(w_1 \cdots w_{N-1}, \bar{x}_N, p) \quad (27)$$

and the short-run profit function is

$$\pi = p \cdot y(w_1 \cdots w_{N-1}, \bar{x}_N, p) - \sum w_k h^k(w_1 \cdots w_{N-1}, \bar{x}_N, p) \equiv \pi(w_1 \cdots w_{N-1}, \bar{x}_N, p) \quad (28)$$

The short-run versions give demands for inputs conditioned on the levels of the fixed inputs. The long-run versions give demands conditioned on the prices of the fixed inputs, assuming that the levels of the fixed inputs are adjusted optimally to maximize profits at those prices.

Up to now we have emphasized the conceptual distinctions among the various types of demand function. What about empirical differences? How much difference is there likely to be among demand elasticities calculated from the different types of demand function? The empirical differences can in fact be very substantial, as we now show. We compare first elasticities calculated from conditional versus unconditional demand functions, then long-run versus short-run demand functions.

The unconditional demand function can be shown to be derived from the conditional demand function by replacing the term $y$ on the right-hand side of (11) with the output supply function (17)—when the firm adjusts its output optimally, the conditional demand function becomes the unconditional demand function. In the long-run context

$$x_k = h^k(w_1 \cdots w_N, p) \equiv g^k[w_1 \cdots w_N, y(w_1 \cdots w_N, p)] \qquad k = 1, \ldots, N-1 \tag{29a}$$

and in the short-run context

$$x_k = h^k(w_1 \cdots w_{N-1}, \bar{x}_N, p)$$
$$\equiv g^k[w_1 \cdots w_{N-1}, \bar{x}_N, y(w_1 \cdots w_{N-1}, \bar{x}_N, p)] \qquad k = 1, \ldots, N-1 \tag{29b}$$

By applying the chain rule of differentiation to (29a) or (29b), one can relate derivatives of the unconditional demand function to those of the conditional demand function. In the long-run case, we obtain the following expressions for the own-price derivative

$$\frac{\partial h^i(w_1 \cdots w_N, p)}{\partial w_i} = \frac{\partial g^i(w_1 \cdots w_N, y)}{\partial w_i} + \frac{\partial g^i(w_1 \cdots w_N, y)}{\partial y} \cdot \frac{\partial y(w_1 \cdots w_N, p)}{\partial w_i} \tag{30}$$

Effect of $\Delta w_i$ on demand for $x_i$ when output can vary | Effect of $\Delta w_i$ on demand for $x_i$ when output is fixed $\leq 0$ | Effect of change in $y$ on demand for $x_i$ | Effect of $\Delta w_i$ on amount of $y$ supplied — $< 0$

the cross-price derivative

$$\frac{\partial h^i(w_1 \cdots w_N, p)}{\partial w_j} = \frac{\partial g^i(w_1 \cdots w_N, y)}{\partial w_j} + \frac{\partial g^i(w_1 \cdots w_N, y)}{\partial y} \cdot \frac{\partial y(w_1 \cdots w_N, p)}{\partial w_j} \tag{31}$$

Effect of $\Delta w_j$ on demand for $x_i$ when output can vary | Effect of $\Delta w_j$ on demand for $x_i$ when output is fixed | Effect of change in $y$ on demand for $x_i$ | Effect of $\Delta w_j$ on amount of $y$ supplied and the output-price derivative

$$\frac{\partial h^i(w_1 \cdots w_N, p)}{\partial p} = \frac{\partial g^i(w_1 \cdots w_N, y)}{\partial y} \cdot \frac{\partial y(w_1 \cdots w_N, y)}{\partial p} \qquad (32)$$

Effect of $\Delta p$ on  Effect of change  Effect of $\Delta p$<br>
demand for $x_i$ when  in $y$ on demand  on amount of<br>
output can vary   for $x_i$   $y$ supplied

Taking the output-price derivative first, the relationship in (32) can be expressed in elasticity terms as

$$\epsilon_p^i = {}_c\epsilon_y^i \cdot \epsilon_p^y \qquad (33)$$

It was noted earlier that the output price elasticity of supply, the second term on the right-hand side of (33), is generally positive. Therefore, the sign of the output-price elasticity of demand depends on whether the input is normal or inferior; the output-price elasticity of demand is positive if the input is normal, and negative if it is inferior.

With the own-price derivative, it turns out that there is an unambiguous difference between unconditional and conditional demand functions. With the unconditional demand, the effect of a change in the price of an input on the demand for that input consists of two components, corresponding to the two terms on the right-hand side of (30). The first component is the effect on input demand when output is held constant—this is the derivative of the conditional demand function. The second component captures impacts resulting from changes in the scale of output. The input price change affects the firm's supply curve, leading it to supply a different level of output. And, the change in output level affects the firm's demand for the input whose price has changed. Converted into elasticities, (30) becomes

$$\epsilon_i^i = {}_c\epsilon_i^i + {}_c\epsilon_y^i \cdot \epsilon_i^y \qquad (34)$$

Using the result noted in (21), it turns out that the two derivatives being multiplied together in the second term on the right-hand side of (34) have opposite signs; therefore the product is negative, regardless of whether the input in question is normal or inferior. This means that the own-price elasticity of the unconditional demand function for an input is negative and larger in absolute value than the own-price elasticity of the conditional demand function. Suppose, for example, that the conditional own-price elasticity of demand is ${}_c\epsilon_i^i = 0.3$, ${}_c\epsilon_y^i = 0.8$, and $\epsilon_i^y = 0.4$; then, the unconditional own-price elasticity is $\epsilon_i^i = 0.62$.

With cross-price derivatives, the elasticity formula corresponding to (31) is

$$\epsilon_j^i = {}_c\epsilon_j^i + {}_c\epsilon_y^i \cdot \epsilon_j^y \qquad (35)$$

The first term on the right-hand side of (35) depends solely on whether inputs $i$ and $j$ are substitutes or complements. If they are substitutes, it is positive; if they are complements it is negative. The more they are substitutes, the more negative it is (i.e., the larger in absolute value); the more they are complements, the more positive it is. The cross-price elasticity of the unconditional demand function differs from this by the second term on the right-hand side of (35), which could be positive or negative. This term is the product of the output elasticity of demand for input $i$ (positive if input $i$ is normal, negative if it is inferior) and the elasticity of output supply with respect to the price of input $j$; as noted previously the latter is opposite in sign to the output elasticity of demand for input $j$ (i.e., negative if input $j$ is normal, positive if it is inferior). Thus, whether the cross-price elasticity of the unconditional demand function is larger or smaller than that of the conditional demand function depends on whether the inputs are normal or inferior. If both inputs are normal and substitutes, the cross-price elasticity of the unconditional demand function is smaller than that of the conditional demand; for example, if $_c\epsilon_j^i = 0.7$, $_c\epsilon_y^i = -0.8$, and $\epsilon_j^y = 0.4$, then $\epsilon_j^i = 0.38$. Conversely, if both inputs are normal and complements, the cross-price elasticity of the unconditional demand function is larger in absolute value than that of the conditional demand; e.g., if $_c\epsilon_j^i = -0.7$, $_c\epsilon_y^i = -0.8$, and $\epsilon_j^y = 0.4$, then $\epsilon_j^i = -1.02$.

For each type of demand function, it can be shown that the long-run version is derived from the short-run version by replacing the term $\bar{x}_N$ on the right-hand side of the short-run demand functions (23) or (26) with the long-run demand function for fixed inputs, as given by (11) or (16)—when the firm adjusts its fixed inputs optimally, the short-run demand functions for variable inputs turn into the long-run demand functions for variable inputs. In the case of conditional demand functions, the relationship is

$$x_k = g^k(w_1, \ldots, w_n, y)$$
$$= g^k(w_1, \ldots, w_{N-1}, g^N(w_1, \ldots, w_n, y), y) \qquad k = 1, \ldots, N-1 \quad (36)$$

with unconditional demand functions, it is

$$x_k = h^k(w_1, \ldots, w_n, p)$$
$$= h^k(w_1, \ldots, w_{N-1}, h^N(w_1, \ldots, w_n, p), p) \qquad k = 1, \ldots, N-1 \quad (37)$$

By applying the chain rule of differentiation to (36) or (37), one obtains expressions relating derivatives of the long-run demand functions to those of the short-run demand functions. For example, from (36) we obtain the following expressions linking the own-price derivatives of long- and short-run conditional demand functions

building. Another major difference is that there tends to be less outdoor space per resident in multifamily units, and the maintenance of the landscaping, including irrigation, may be controlled by a building manager rather than by the residents individually. Demographically, multifamily units are often associated with smaller family sizes, although the opposite may be true in lower-income areas where large households may live in apartment units because they cannot afford to buy a home of their own. Also, multifamily units may have somewhat fewer water-using appliances than single-family units. The nonmetering of individual dwelling units within multifamily structures reduces the incentive to fix leaks or avoid waste; however, the other differences all tend to reduce per capita use in multifamily versus single-family residences. Overall, the net effect is generally lower per capita use in multifamily units. In the MWD service area, for example, residential water use in a normal year area is now estimated to average about 150 gallons per capita per day (GPCD) for single-family units and 110 GPCD for multifamily units.

Table 2-3 presents data on the breakdown of residential water use among different end uses for single- versus multifamily units in the MWD service area. It should be emphasized that these figures are estimates

**Table 2-3.** Breakdown of Water Use in Single- and Multifamily Residences in MWD Service Area

| Water use category | Water use (gallons per capita per day) | |
| --- | --- | --- |
| | Single-family | Multifamily |
| Indoor: | | |
| Toilets | 30 | 30 |
| Showers/bath | 27 | 25 |
| Washing clothes | 21 | 17 |
| Cooking/cleaning | 13 | 13 |
| Dishwashing | 6 | 4 |
| Subtotal | 97 | 89 |
| Outdoor: | | |
| Landscape irrigation, gardening | 46 | 18 |
| Cooling | | 1 |
| Swimming pool, car washing, and other outdoor uses | 7 | 2 |
| Subtotal | 53 | 21 |
| Total | 150 | 110 |

SOURCE: Urban Water Use Characteristics in the Metropolitan Water District of Southern California, Metropolitan Water District, Planning Division, April 1993, Tables 2-5 and 2-6.

based on expert judgment rather than actual field measurement. Unlike electricity where considerable effort has been devoted to measuring appliance ownership and actual patterns of usage for different purposes within the home, there has been relatively little monitoring of water-related appliance ownership or water-usage patterns within individual homes. There is a wealth of engineering information on water usage by specific appliances under theoretical operating conditions. For example, conventional toilets use about 5.5 gallons per flush. Low-flush toilets mandated in California and other states around 1980 use 3.5 gallons per flush. In California, this law was revised in 1993 to require ultra-low-flush toilets that use 1.5 gallons per flush. But, what is not known is how many times per day a person flushes the toilet, whether people flush more often with ultra-low-flush compared to conventional low-flush toilets, whether they modify their toilets (e.g., by placing a brick in the toilet bowl), or whether toilets leak and how people deal with this. In short, there is a behavioral component in addition to the engineering component of indoor water use, and this is not yet well understood.

As an illustration, during much of the 1980s water planners in California and elsewhere generally assumed that indoor water use averaged about 77 GPCD in a traditional nonconserving single-family home, and about 60 GPCD in a conserving home, based on information about water savings from conservation developed by Brown and Caldwell (1984).* Depending on the fraction of the housing stock built after the requirement for low-flush toilets had gone into effect, planners typically assumed that overall single-family indoor use was some weighted average of 60 and 77 GPCD. This is now considered too low in California; as Table 2-3 shows, MWD now puts indoor single-family residential use at 97 GPCD. This is despite the gains from low-flush toilets and other conservation measures introduced during the 1980s and early 1990s. The explanation seems to be that, although there were some real gains from conservation involving indoor residential water use, there were also some changes in appliance characteristics and in appliance ownership and usage that may have offset these gains. For example, whereas the typical capacity of a bath tub in a home was 50 gallons, some people are now installing tubs with capacities of 80 to 100 gallons. In consequence,

---

* The Brown and Caldwell study assumed that people flush toilets at home about 4 times per person per day, and low-flush toilets use 3.5 gallons per flush versus 5.5 gallons per flush for traditional toilets. However, it also allowed for some increased leakage from low-flush toilets. It assumed that low-flow showerheads use about 2 instead of 3.5 gallons per minute for about 4.8 minutes per person per day; conserving dishwashers use 8.5 instead of 14 gallons per load, with 0.17 loads per capita per day; and conserving clothes washing machines use 42 instead of 55 gallons per load, with 0.3 loads per capita per day.

there remains considerable uncertainty regarding the actual nature of indoor water use.*

While indoor residential water usage depends crucially on the types of appliances owned and how these are used, outdoor residential water usage depends crucially on lot size, climate, and the style of landscaping. For a given type of groundcover, irrigation requirements per square foot can vary greatly depending on the type of soil, the slope of the land, the amount and timing of precipitation, temperature, wind, and other factors. The type of landscaping also has a substantial effect, for example turf versus xeriscape. The type of turf itself can make a difference; warm-season turfgrasses such as bermuda grass require about 20 percent less water in California than cool-season turfgrasses such as Kentucky bluegrass. Moreover, in addition to the physical considerations associated with soil type and plant type, as with indoor water use there is an essential behavioral component in outdoor water use. People choose to adopt one style of landscaping rather than another, and they water their landscape with different degrees of knowledge, care, and attentiveness. A survey of 515 single-family residential landscapes in the MWD service area in 1990 found that only 11 percent of the households irrigated at levels within ±10 percent of what was required, given the size of their landscaped area and the type of ground cover. About 39 percent of the households overirrigated in the sense of applying at least 10 percent more water per unit area than was required by the type of ground cover, while 50 percent underirrigated, in the sense of applying at least 10 percent less than the required irrigation (Kiefer and Dziegielewski 1991). Lot size was an important factor here: households with landscape areas of less than 3,500 square feet all overirrigated, while those with landscape areas greater than 8000 square feet all underirrigated.

Although outdoor use accounts for about only one-third of total residential use, it is more important than indoor use as a source of variability across space and time. Indoor water use in single-family residences is generally likely to be found somewhere in the range of, say, 70–110 GPCD; outdoor use in single-family residences can range anywhere from less than 30 GPCD to over 100 GPCD. Some indication of the variability in outdoor use can be seen by the comparison of total residential use across the five main counties in the MWD service area shown in Table 2-4. Riverside and San Bernadino are both interior counties; the other three counties all have segments with a cooler, coastal climate. Riverside is the hottest and most arid county within the MWD service area, and water use in single-

---

* For example, a 1991 report estimates total residential use in the MWD service area in 1985 as 140 GPCD for single-family units and 94 GPCD for multifamily units, compared to the later figures of 150 and 110 GPCD reported in Table 2-3 (Dziegielewski and Opitz 1991, Table II-3).

**Table 2-4.** Geographic Variation in Residential Water Use, MWD Service Area 1990

| County | Water use per unit (gallons per unit per day) | |
| --- | --- | --- |
| | Single-family units | Multifamily units |
| Los Angeles | 451 | 256 |
| Orange | 520 | 301 |
| Riverside | 554 | 299 |
| San Bernardino | 519 | 342 |
| San Diego | 417 | 273 |

SOURCE: Dziegielewski and Opitz (1991), Appendix F.

family dwellings there averages about 100 gallons per day more per unit than in Los Angeles county.

## Economic Theory of Consumer Demand for Water

As with industrial water use, analysts have approached the forecasting of residential use in two ways. The water requirements approach uses fixed coefficients for residential water use per capita, perhaps stratified by dwelling type—$x$ GPCD for single-family units and $y$ GPCD for multifamily units, where $x$ and $y$ are based on data like that given in Table 2-3—which are multiplied by forecasts of future population in the water agency's service area. The economic approach treats per capita water use as a behavioral phenomenon, reflected in a consumer-demand function for water that can be estimated from data on actual consumption. This section summarizes the economic theory of a consumer's demand for a final good as applied to residential water use. Because of the tendency not to meter individual units in multifamily dwellings, the main application of this theory will be to single-family residential use which generally is metered.

The theory focuses on the optimizing behavior of a consumer faced with a limited budget. The consumer could be either a single individual or a group of individuals like a household that share a common budget and allocate it collectively. Here we will treat the consumer as a single individual, but that is merely intended to simplify the exposition; it could apply equally well to decision making by a household. The starting point of the theory is the concept of the consumer's utility function, which characterizes her level of well-being $u$ as a function of what she consumes

$$u = f(x_1, \ldots, x_N) \tag{46}$$

where there are $N$ commodities and $x_i$ denotes the amount of the $i$th commodity consumed per unit of time, $I = 1, \ldots, N$. For each possible consumption pattern, represented by some particular combination of $x$s, the utility function states the corresponding level of well-being, measured on what is known as an ordinal scale. The information contained in the utility function may be conveyed in the form of an algebraic equation, a graph, or a table. One particular example is the Cobb-Douglas utility function, analogous to (8) with

$$u = Ax_1^{\alpha_1} x_2^{\alpha_2} \cdots x_N^{\alpha_N} \tag{47}$$

where $A > 0$ and $\alpha_i > 0$, $I = 1, \ldots, N$, are coefficients whose numerical values would be estimated statistically from data on consumption behavior, with $\sum \alpha_i = 1$. A variant of this is what is known as the Stone-Geary utility function which takes the form

$$u = A(x_1 - \gamma_1)^{\alpha_1} (x_2 - \gamma_2)^{\alpha_2} \cdots (x_N - \gamma_N)^{\alpha_N} \tag{48}$$

where the $\gamma$s are interpreted as minimum levels of consumption—that is, one gets utility from a commodity only to the extent that its consumption exceeds $\gamma_i$.

The role of water in the individual's preferences can be treated in two ways. One approach lumps all the different uses of water, both indoor and outdoor, into a single commodity and models the individual as having a demand for water as a whole. The other approach disaggregates the demand for water into the separate end uses (e.g., clothes washing, lawn watering, etc.) and treats each as a separate commodity for which the individual has a separate demand. Because of the paucity of data on residential end uses of water, almost all of the existing literature adopts the first approach, treating water as a single, lumped commodity. We will therefore focus mainly on this approach. Following this approach, let $x_1$ denote the total amount of water used by the individual (i.e., overall per capita residential water use); let $x_2, \ldots, x_N$ denote consumption of goods other than water. Let $p_1$ be the per unit charge for water (e.g., in units of dollars per hundred cubic foot per month), and let $p_2, \ldots, p_N$ be the prices of the other goods. The individual's income available to be spent on consumption is denoted $y$. Total expenditure cannot exceed this income, a restriction known as the *budget constraint:*

$$p_1 x_1 + p_2 x_2 + \cdots + p_N x_N = y \tag{49}$$

Whereas there are two possible types of demand functions for an input that come out of the economic theory of producer behavior, there is just one type of demand function for a final good that comes out of the economic theory of consumer behavior. The individual is assumed to choose what to consume so as to maximize her utility subject to the budget constraint, choosing $x_1, \ldots, x_N$ so as to

$$\text{maximize} \quad u(x_1, \ldots, x_N) \quad \text{subject to} \sum_i p_i x_i = y \quad (50)$$

The solutions for $x_1, \ldots, x_N$ are the optimal quantities the individual should purchase when she has an income level of $y$ and faces commodity prices $p_1, \ldots, p_N$. The behavioral rule relating the givens in the consumer's problem to her optimal choice of consumption levels is a set of functions of the form

$$x_i = h^i(p_1, \ldots, p_N, y) \quad i = 1, \ldots, N \quad (51)$$

which are known as the consumer's *ordinary* or *Marshallian demand functions*. For example, with the Cobb-Douglas utility function in (47) the ordinary demand functions take the form

$$x_i = \frac{\alpha_i \cdot y}{p_i} \quad i = 1, \ldots, N \quad (52)$$

whereas with the Stone-Geary utility function in (48), the ordinary demand functions are

$$x_i = \gamma_i + \frac{\alpha_i}{p_i} \left( y - \sum_{j=1}^{N} p_j \gamma_j \right) \quad i = 1, \ldots, N \quad (53)$$

For example, when $N = 2$, the Stone-Geary demand function for good 1 is

$$x_1 = (1 - \alpha_1)\,\gamma_1 + \alpha_1 \frac{y}{p_1} - \alpha_1 \gamma_2 \frac{p_2}{p_1} \quad (53')$$

The derivatives of the ordinary demand functions describe the sensitivity of the consumer's demand for a commodity to a change in its own price, the prices of other goods, and her total income. As with the firm's demand for inputs, these sensitivities can be expressed in percentage terms as elasticities that measure the proportional change in consumer demand per 1 percent change in prices or income. The own-price elasticity of demand is

$$\eta_i^i \equiv \frac{p_i}{x_i} \frac{\partial h^i(p_1, \ldots, p_N, y)}{\partial p_i}$$

the cross-price elasticity of demand is

$$\eta_j^i \equiv \frac{p_j}{x_i} \frac{\partial h^i(p_1, \ldots, p_N, y)}{\partial p_j}$$

and the income elasticity of demand is

$$\eta_y^i \equiv \frac{y}{x_i} \frac{\partial h^i(p_1, \dots, p_N, y)}{\partial y}.$$

With the demand functions in (52) from a Cobb-Douglas utility function, for example, $\eta_i^i = -1$, $\eta_j^i = 0$, and $\eta_y^i = 1$. However, although the Stone-Geary utility function is superficially similar to the Cobb-Douglas utility function, the demand elasticities associated with (53) differ considerably from those associated with (52).

In general, the demand elasticities can have either sign. When the income elasticity is positive, the commodity is said to be a *normal* good; when it is negative, the commodity is said to be an *inferior* good. If the elasticity is positive and greater than unity, the commodity is said to be a *luxury*, since consumption of the good will be disproportionately higher when the consumer is richer. For all normal goods, the own-price elasticity is negative—that is, a higher price for a good leads to a fall in its consumption; only for strongly inferior goods can the own-price elasticity be positive. It can safely be assumed that for most uses, the residential demand for water is a normal good and the demand curve slopes down. If the own-price elasticity is negative but less than unity in absolute value, the demand is said to be *inelastic;* if it exceeds unity in absolute value, the demand is said to be *elastic.* The results on own-price elasticities of input demand functions and expenditure mentioned in footnote 5 also apply to the consumer's ordinary demand function: if the demand for a commodity is inelastic, an increase in its price leads to an increase in total expenditure on the commodity, and conversely for a price reduction. As discussed in the next section, the evidence from the United States generally suggests that the residential demand for indoor water is inelastic, while the residential demand for outdoor use may be elastic. If the cross-price elasticity is negative (e.g., an increase in $p_j$ lowers the demand for $x_i$), the two goods are considered complements; if the cross-price elasticity is positive, they are considered substitutes. For example, with electric appliances such as dishwashers, electricity is a complement to water. But, electricity is a substitute for water in space cooling, where electric air conditioners compete with the older technology known (not altogether affectionately) as swamp coolers.*

Just as with production functions, there are alternatives to the Cobb-Douglas, Linear Expenditure System, and related utility functions which are more flexible and are based on the newer approach that uses duality theory. For example, there is a translog utility function which is a counter-

---

* Swamp coolers are usually classified as outdoor water use, and are treated this way in Table 2-3 under multifamily units.

part to the translog cost function in (43). These flexible forms have not yet been used much in modeling the residential demand for water.

Instead, the literature on residential demand has frequently taken the more ad hoc approach of estimating a single-demand equation for water using some simple functional form such as a linear demand equation in which the residential demand for water (denoted by $x_1$, say) is regressed on the price of water and consumer income ($y$):

$$x_1 = \alpha - \beta p_1 + \gamma y \tag{54}$$

In addition, the researcher might include a variable for the price of some goods that are important complements or substitutes for water, $p_2$:

$$x_1 = \alpha - \beta p_1 + \gamma y + \delta p_2 \tag{55}$$

Commonly used alternatives to the linear demand equation in (54) are the log-log model, which is linear in the logarithms of price and quantity

$$\ln x_1 = \alpha - \beta \ln p_1 + \gamma \ln y \tag{56}$$

and various semilog models in which log quantity is regressed on price

$$\ln x_1 = \alpha - \beta p_1 + \gamma y \tag{57}$$

or quantity is regressed on log-price*

$$x_1 = \alpha - \beta \ln p_1 + \gamma \ln y \tag{58}$$

The own-price and income elasticities associated with these demand equations are as follows:

|                        | $\eta_1^1$        | $\eta_y^1$        |
|------------------------|-------------------|-------------------|
| linear demand (54)     | $-\beta p_1 / x_1$ | $\gamma y / x_1$  |
| log-log demand (56)    | $-\beta$          | $\gamma$          |
| semilog demand (57)    | $-\beta p_1$      | $\gamma y$        |
| semilog demand (58)    | $-\beta / x_1$    | $\gamma / x_1$    |

While these four demand equations have an advantage of simplicity, they are formally inconsistent with economic theory because, as they stand, they cannot be derived from the maximization of a utility function along the lines of (50). For this reason, they are rejected by Al-Qunaibet

---

* Al-Qunaibet and Johnston (1985), who apply these models to the municipal demand for water in Kuwait, call (58) the semilog model and refer to (57) as the price-exponential model.

and Johnston (1985) in favor of utility-based demand functions for water such as the Cobb-Douglas (52), and the Linear Expenditure System (53). However, if one makes a small but economically significant modification, demand equations such as (54) and (56)–(58) can in fact be rendered consistent with utility maximization. The modification involves the assumption that the commodities other than water can be lumped into a single aggregate, so that the consumer is conceived as having preferences for just two items (I) water and (ii) all other commodities as a group; that is, his utility function is $u = u(x_1, x_2)$, where $x_1$ is the consumption of water and $x_2$ is the consumption of all other commodities. In that case, it is possible to obtain a demand for water like the linear, log-log, or semilog demand functions. In the linear case, the demand function technically has the form

$$x_1 = \alpha - \beta \frac{p_1}{p_2} + \gamma \frac{y}{p_2} \qquad (59)$$

which is not the same as (55) because $p_2$ enters as the denominator in ratios with $p_1$ and $y$ rather than as a separate variable. The economic logic is that it is not the absolute levels of prices and income that matter to a consumer who solves (50) but rather relative prices and income; in (59), these are expressed relative to the price of all nonwater goods, $p_2$. But, it is always possible to normalize prices so that $p_2 = 1$, in which case (59) coincides with (54). Similarly (56), (57), and (58) can always be construed to be functions of $(p_1/p_2)$ and $(y/p_2)$ with prices normalized so that $p_2 = 1$. Under those circumstances, the linear, log-log, and semilog demand equations are as consistent with economic theory as the Cobb-Douglas and Stone-Geary demand equations in (52) and (53).

Although all potentially consistent with economic theory, these demand equations can produce empirical results that are very different, as Al-Qunaibet and Johnston (1985) found when fitting alternative-demand models to data on municipal water use in Kuwait. The resulting estimates of the own-price elasticity of demand $\eta_1^1$ ranged from $-0.771$ for the Stone-Geary demand equation (53) to $-0.861$ for the linear demand equation (54) and $-0.957$ for the semilog equation (57); the estimates of the income elasticity $\eta_y^1$ ranged from 0.01 for the semilog to 0.022 for the linear and 0.211 for the Stone-Geary.

Which demand equation should one use? One criterion for selecting an equation is goodness of fit—which of the alternative equations best fits the data according to some statistical measure such as the coefficient of determination or the F-test. Another criterion that should also be considered, however, is the plausibility of the qualitative properties associated with the alternative-demand functions. One qualitative property involves the price or income elasticities: these are constant with the log-log demand equation, but increase in absolute value with price or income, in the case of the semilog (57), or inversely with the quantity demanded in the case of (58).* Another qualitative property involves the limiting behavior of demand as price rises. With both the linear demand equation (54) and the semilog (58), the quantity demanded falls to zero when price reaches a certain, finite level. With the log-log (56) and semilog (57), by contrast, demand falls to zero only asymptotically as price rises to infinity. And, with the Stone-Geary (53), as price rises demand falls asymptotically to $\gamma_1$ rather than to zero. To the extent that some minimum level of water use is a necessity for most households, this is a more plausible notion. However, the same effect can be induced in the log-log and semilog equations by modifying them slightly. For example, the modified version of (57) is

$$\ln (x_1 - \theta_1) = \alpha - \beta p_1 + \gamma y \tag{60}$$

or, equivalently,

$$x_1 = \theta_1 + \exp (\alpha - \beta p_1 + \gamma y)$$

where $\theta_1$ plays the same role as $\gamma_1$ in the Stone-Geary model.[†]

In addition to the choice of functional form, another issue facing the researcher is the choice of explanatory variables to include in the demand function along with the economic variables of price and income. One should include whatever factors influence the consumer's demand, including demographic variables (e.g., sex, age, size, and composition of household), physical characteristics (e.g., age of house, number of bathrooms, size of lot, style of landscaping), and climatic conditions (e.g., temperature, rainfall).[‡] Denote these variables by $z$. The logic is that two

---

* See Griffin and Chang (1991) for an example of a study which takes the plausibility of the estimates of demand elasticities as the criterion for comparing alternative forms of demand equation. However, they apply different names to the demand equations than are used here. For example, they apply the Cobb-Douglas label to the log-log demand equation (56) rather than (52). We use labels such as Cobb-Douglas or translog that characterize the form of the underlying utility function, while they use labels that characterize the form of the demand equation itself.

[†] To be consistent with utility theory, the demand equation is construed as a function of $(p_1/p_2)$ and $(y/p_2)$ with $p_2 = 1$.

[‡] While household size and composition clearly affect the residential demand for water, the elasticity of household demand with respect to household size appears to be significantly less than unity—that is, water use per capita falls as household size increases. Most existing studies find an elasticity of about 0.4 to 0.6; for a summary, see Table A-5 in Dziegielewski and Opitz (1991). If there are typically about four people per household and single-family residential use averages 150 GPCD, this implies that, at the margin, residential use increases by 60–90 gallons per day for each additional household member.

people with a different $z$ would obtain a different utility from the same $x$s. Therefore, $z$ enters the utility function along with the $x$s: $u = u(x_1, \ldots, x_N, z)$. Through the maximization of utility as in (50), the demand functions for the $x$s also come to depend on $z$; in place of (51), we have $x_i = h^i(p_1, \ldots, p_N, y, z)$, $I = 1, \ldots, N$. In practice, the introduction of $z$ into demand functions such as (54) and (56)–(60) can be implemented by making one or more of their coefficients functions of $z$.

In the literature, this approach has frequently been used to allow for the effects of climate on residential water demand. For example, Al-Qunaibet and Johnston (1985) introduce weather conditions into their Stone-Geary demand function by making the subsistence parameter for water, $\gamma_1$, a function of temperature $T$, minutes of sunshine $S$, and wind speed $W$:

$$\gamma_1 = \kappa + \lambda \ln T + \mu \ln S + \nu \ln W \qquad (61)$$

where $\kappa$, $\lambda$, $\mu$, and $\nu$ are coefficients to be estimated along with $\alpha_1$, $\alpha_2$, and $\gamma_2$. Similarly, researchers using the linear, log-log, or semilog demand equations have made the intercept $\alpha$ a function of climate variables.[*] The same could be done with $\beta$, to allow for differences in the elasticity of demand for indoor versus outdoor uses. As an alternative to the explicit inclusion of climate variables in the demand equation, another way to capture the effect of seasonality on residential demand for water is to allow the coefficients of the demand equation to vary by season, adding a dummy variable for winter versus summer which shifts the intercept $\alpha$, or the price coefficient $\beta$.[†] Going beyond this, one can estimate separate demand equations for different seasons—for example, estimating one equation for winter months and another for summer months—which effectively allows all the coefficients to vary with the season.[‡] A related approach is to estimate one equation for winter-period demand viewed as base use or weather-insensitive use, that is, primarily indoor use, and a separate equation for seasonal use, which is weather-sensitive and is calculated as the difference between base use and total use during the summer months of the year.[§]

---

[*] See, for example, Anderson et al. (1980) and Carver and Boland (1980); a sophisticated recent analysis is Chesnutt, Bamezai, and McSpadden (1992). When the stochastic error term in the demand equation is made a function of climate variables, this can generate autoregressive or other time-series models of water use; for examples, see Maidment and Miaou (1986) and Smith (1988).

[†] See Griffin and Chang (1991) for an example.

[‡] See Howe and Linaweaver (1967) and Howe (1982) for an example.

[§] See Maidment and Miaou (1986) for an example.

We have focused so far on an approach which aggregates all the different residential uses of water into a single commodity. The alternative, disaggregated approach treats different uses as separate commodities, each with its own demand equation. Apart from requiring multiple-demand equations, the disaggregated approach may not differ much from the aggregated approach. Often, however, there can be some significant differences which arise from the fact that certain end uses are associated with the ownership of particular appliances: if the appliance is not owned, there is no end use. Therefore, the demand for the end use is intertwined with a demand for ownership of the appliance. Moreover, ownership of the appliance is often what is known as a *discrete* rather than *continuous* choice—that is, ownership is characterized through a discrete variable (owns one versus doesn't own one) rather than a continuous variable (how many units are owned). The economics and econometrics of discrete choices lie beyond the scope of this chapter; here we offer only a brief sketch.

With the disaggregated approach, the demand for water in an appliance-related end use can be decomposed into three factors

$$
\begin{matrix}
\text{Amount of water} & \text{The number} & \text{extent to which} & \text{amount of water} \\
\text{demand in connection} = \text{of appliances} & \cdot & \text{the appliances are} & \cdot & \text{consumed each time} & (62) \\
\text{with the use of a type} & \text{owned} & \text{used} & \text{the appliances} \\
\text{of appliance} & & & \text{are used}
\end{matrix}
$$

Let $x_k$ denote the amount of water demanded by an individual or a household in connection with the use of the $k$th type of appliance, $k = 1, \ldots, K$; let $n_k$ be the number of this type of appliances that the individual or household owns;* let $u_k$ denote the level of usage of these appliances (e.g., number of times per day or number of hours per week), and let $w_k$ be the amount of water consumed per unit of application. Then the decomposition in (62) becomes

$$
x_k = n_k \cdot u_k \cdot w_k \tag{62'}
$$

It would be natural for $w_k$ to be a fixed constant for the individual, determined by the engineering design of the appliance (e.g., whenever a dishwasher is run, it uses $w_k = 20$ gallons per application). However, $n_k$ and $u_k$ are variables representing choices by the individual. The individual can

---

* If this is a situation where ownership is essentially a discrete choice, so that the main question is whether the individual owns the item rather than how many units one owns, $n_k$ becomes a *binary* variable taking the value $n_k = 1$ if the individual does own the item and $n_k = 0$ if he or she doesn't. The expected value of $n_k$ is then the proportion of the population who own the item.

be thought of as having a utility function defined over $n_k$ and $u_k$, as well as the consumption of other market commodities. In effect, $n_k$ and $u_k$ are separate commodities, each with its own price and its own demand equation. The price associated with $n_k$ is the capital cost of owning the appliance denoted, say, by $\pi_k$; the price associated with $u_k$ is the operating cost of running the appliance denoted, say, by $c_k$. The operating cost depends partly on the price of water, say $p_w$, and partly on other prices, such as the price of electricity, so that $c_k = w_k p_w + c^*$, where $c^*$ denotes the nonwater component of operating expenses. Let $p$ denote the vector of prices of the other market commodities and $y$ the consumer's income. The consumer chooses the optimal number of appliances to own and their optimal level of usage, along with the consumption of other commodities, in a utility maximization like (50). The resulting ordinary demand functions are all functions of $p, y, \pi_1, \ldots, \pi_K$, and $c_1, \ldots, c_K$. Hence, when there is a change in the price of water, $p_w$, the resulting change in demand for the $k$th end use, expressed in elasticity form, is

$$\eta_k = (\eta_n^k + \eta_u^k) \cdot \eta_c^k \tag{63}$$

where

$$\eta_k \equiv \frac{\partial x_k}{\partial p_w} \cdot \frac{p_w}{x_k} \qquad \eta_n^k \equiv \frac{\partial n_k}{\partial c_k} \cdot \frac{c_k}{n_k}$$

$$\eta_u^k \equiv \frac{\partial u_k}{\partial c_k} \cdot \frac{c_k}{u_k} \qquad \eta_c^k \equiv \frac{\partial c_k}{\partial p_w} \cdot \frac{p_w}{c_k}$$

The elasticity $\eta_c^k$ measures the degree to which a change in $p_w$ raises the overall operating cost of the appliance. The elasticities $\eta_n^k$ and $\eta_u^k$ measure, respectively, how a change in the operating cost affects the ownership and usage of the appliance. While there is little information on these elasticities for water, the experience with electricity prices suggests that usage is often price-inelastic but ownership may ultimately be quite price-elastic, so that the magnitude of $\eta^k$ may be determined largely by $\eta_u^k$.

## Empirical Estimates
## of M&I Demand Elasticities

In this chapter we have focused on the economic theory of M&I use, treating water as either an input to production or a final good. In both cases, the theory is cast in terms of decision making by an individual agent, whether a firm or a household, looking at a specific use of water viewed as an argument in a production function or a utility function. This approach leads to demand functions by individual producers or consumers for specific uses of water. By contrast, most of the readily available data on M&I use tends to be aggregated—aggregated over individual users and aggregated over distinct uses of water. For example, there is data on total single-family residential use within a city, or total residential use, or total industrial use, or just total M&I use. Therefore, empirical analyses of real-world data have often involved something of a leap from a theory that applies to individual agents and distinct uses to data that are more aggregate than the theory contemplates. Since residential use is typically the largest component of M&I use, as illustrated in Table 2-1, researchers have tended to employ the economic theory of consumer demand as their paradigm for the analysis of M&I use, using demand functions such as the linear (54), log-log (56) or semilog (57). The results of these analyses are usually summarized in terms of the own-price and income elasticities of demand, $\eta_1^1$ and $\eta_y^1$ in the notation used previously. Table 2-5 summarizes the estimates of demand elasticities from the literature on M&I use in the United States, noting where appropriate the sensitivity of the estimate to the type of data and the choice of functional form.

# References

Agthe, Donald E. and R. Bruce Billings. 1980. Dynamic Models of Residential Water Demand. *Water Resources Research*, 16(3): 476–480.

Al-Qunaibet, Mohammad H, and Richard S. Johnston. 1985. Municipal Demand for Water in Kuwait: Methodological Issues and Empirical Results *Water Resources Research*, 2(4): 433–438.

Anderson, R. L., T. A. Miller, and M. C. Washburn. 1980. Water Saving from Lawn Watering Restrictions During a Drought Year. *Water Resources Bulletin*, 16(4): 642–645.

Andrews, A. R. and K. C. Gibbs. 1975. An Analysis of the Effect of Price on Residential Water Demand: Metropolitan Miami, Florida. *Southern Journal of Agricultural Economics*, 7(1): 125–130.

Babin, Frederick, Cleve Willis, and Geoffrey Allen. Estimation of Substitution Possibilities Between Water and Other Production Inputs. *American Journal of Agricultural Economics*. February 1982.

Bain, J. S., R. E. Caves, and J. S. Margolis. 1966. *Northern California's Water Industry*. Baltimore, MD: Johns Hopkins University Press.

Ben-Zvi, Samuel. 1980. *Estimates of Price and Income Elasticities of Demand for Water in Industrial Use in the Red River Basin*. Tulsa, OK: U.S. Army Corps of Engineers.

Berndt, Ernst R. 1991. *The Practice of Econometrics: Classic and Contemporary*. Reading, MA: Addison-Wesley Publishing.

Billings, R. Bruce. 1982. Specification of Block Rate Price Variables in Demand Models. *Land Economics*, 58(3): 386–394.

Billings, R. B. and D. E. Agthe. 1980. Price Elasticities for Water: A Case of Increasing Block Rates. *Land Economics*, 56(1): 73–84.

**Table 2-5.** Price and Income Elasticities in North America

| Date of publication | Authors | Study location | Type of use | Type of data | Type of model | Own-price elasticity | Income elasticity |
|---|---|---|---|---|---|---|---|
| 1951 | Larson and Hudson | 15 communities in Illinois | M&I | CS | L | | 0.7 |
| 1957 | Seidel and Baumann | 111 areas in the U.S. | M&I | CS | LL | -0.12 to -1.00 | |
| 1958 | Fourt | 34 U.S. cities | M&I | CS | LL | -0.39 | 0.28 |
| 1963 | Bain | Northern California | M&I | CS | LL | -1.1 | 0.58 |
| 1963 | Gottlieb | Kansas | M&I | CS | LL | -0.67 to -1.23 | 0.28 |
| 1963 | Headley | San Francisco Bay Area | M&I | TS | L | | 0.4 |
| 1964 | Gardner-Schiek | Northern Utah | M&I | CS | LL | -0.77 | |
| | | Northern Utah | M&I | CS | L | -0.67 | |
| 1965 | Flack | | M&I | CS | L | -0.12 to -0.61 | |
| 1966 | Ware and North | Georgia households | SF Residential | CS | LL | -0.61 | 0.38 |
| | | Georgia households | SF Residential | CS | L | -0.67 | 0.83 |
| 1967 | Conley | So. California | Agg. Residential | CS | | -1.02 | |
| 1967 | Howe and Linaweaver | 21 areas across the U.S. | SF Residential, annual | CS | L | -0.4 | 0.47 |
| | | 21 areas across the U.S. | SF Residential, winter | CS | L | -0.23 | 0.32 |
| | | 11 areas in the east | SF Residential, summer | CS | L | -1.57 | 1.45 |
| | | 11 areas in the east | | | | | |
| | | 10 areas in the west | SF Residential, summer | CS | L | -0.73 | 0.69 |
| | | 10 areas in the west | | | | | |
| 1969 | Turnovsky | 19 Massachusetts towns | Agg. Residential | CS | L | -0.25, -0.28 | |
| 1970 | Male, Willis, Babin, and Shilito | Northeast US | Agg. Residential | CS | L | -0.2 | |
| 1970 | Hittman Assoc. | Northeast US | Agg. Residential | CS | LL | -0.37 | |
| 1971 | Young | United States | Agg. Residential | CS | LL | -0.44 | |
| | | Tucson, Arizona | M&I | TS | L | -0.65 | |
| | | Tucson, Arizona | M&I | TS | LL | -0.69 | |
| 1972 | Wong | City of Chicago, 1951–61 | M&I | TS | LL | -0.02 | 0.2 |
| | | Chicago suburbs, 1951–61 | M&I | TS | LL | -0.28 | 0.26 |

**Table 2-5.** Price and Income Elasticities in North America (*Continued*)

| Date of publication | Authors | Study location | Type of use | Type of data | Type of model | Own-price elasticity | Income elasticity |
|---|---|---|---|---|---|---|---|
| | | Chicago suburbs, over 25,000, 1961 | M&I | CS | LL | -0.53 | 1.03 |
| | | Chicago suburbs, 10–25,000, 1961 | M&I | CS | LL | -0.82 | 0.84 |
| | | Chicago suburbs, 5–10,000, 1961 | M&I | CS | LL | -0.46 | 0.48 |
| | | Chicago suburbs, under 5,000, 1961 | M&I | CS | LL | -0.27 | 0.58 |
| 1972 | Grima | Toronto, Ontario | SF Residential, annual use | CS | LL | -0.93 | 0.56 |
| | | Toronto, Ontario | SF Residential, winter use | CS | LL | -0.75 | 0.41 |
| | | Toronto, Ontario | SF Residential, summer use | CS | LL | -1.07 | 0.51 |
| 1973 | Morgan | 92 residences in Santa Barbara County | SF Residential, winter use | CS+TS, monthly | L | -0.49 | 0.53 |
| | | 92 residences in Santa Barbara County | SF Residential, winter use | CS+TS, monthly | LL | | 0.43 |
| 1974 | Primeaux and Hollman | 402 households in 14 Mississippi cities | SF Residential | CS | LL | -0.45 | 0.24 |
| | | 402 households in 14 Mississippi cities | SF Residential | CS | L | -0.37 | 0.26 |
| 1974 | Sewell and Roueche | 17 areas in Victoria, B.C., 1954–70 | M&I | CS+TS | L | -0.46 | 0.27 |
| | | 17 areas in Victoria, B.C., 1954–70 | M&I | CS+TS | LL | -0.39 | 0.19 |
| 1975 | Andrews and Gibbs | Miami, Florida | M&I | CS | SL | -0.62 | 0.8 |
| 1975 | Hogarty and Mackay | | SF Residential, short-run | CS+TS | L | -0.86 | |

| Year | Author | Sample | Use | Data | Run | Elasticity | Income |
|---|---|---|---|---|---|---|---|
| 1975 | Pope, Steppl, Lytle | South Carolina households, 1965–71 | SF Residential, long-run | CS+TS | L | −0.56 | |
| | | South Carolina households, 1965–71 | SF Residential, irrigators | CS+TS | L | −0.31 to −0.67 | |
| | | | SF Residential, non-irrigators | CS+TS | L | −0.06 to −0.36 | |
| 1976 | Grunewald, Haan, Debertin, and Carey | Kentucky | Agg Residential | CS | LL | −0.92 | |
| 1976 | Morgan and Smolen | 33 areas in Southern California | M&I, annual | CS | L | −0.44 | 0.33 |
| | | 33 areas in Southern California | M&I, winter | CS | L | −0.45 | |
| | | 33 areas in Southern California | M&I, summer | CS | L | −0.43 | |
| 1977 | Clark and Goddard | 22 areas | M&I | CS | L | −0.63 | |
| 1977 | | 22 areas | M&I | CS | LL | −0.6 | |
| 1977 | Gallagher and Robinson | | SF Residential, winter | CS+TS | LL | −0.24 | |
| 1977 | Gardner | Minnesota areas | Agg Residential | CS | L | −0.24 | |
| | | Minnesota areas | Agg Residential | CS | LL | −0.15 | |
| 1977 | Danielson | Raleigh, North Carolina | M&I, annual | TS | | −0.27 | |
| | | Raleigh, North Carolina | M&I, winter | TS | | −0.305 | |
| | | Raleigh, North Carolina | M&I, summer | TS | | −1.38 | |
| 1978 | Camp | 288 households in 10 Mississippi cities | SF Residential | CS | L | −0.24 | |
| 1978 | Carver | Fairfax, Co, 1974–75 | Agg Residential, summer | TS, monthly | L | −0.13 to −0.17 | |
| | | Fairfax, Co, 1974–75 | Agg Residential, winter | TS, monthly | L | −0.02 to −0.04 | |
| 1978 | Gibbs | Households in Miami, Florida | SF Residential | CS+TS | L | −0.51 to −0.61 | 0.51 to 0.8 |
| 1979 | Cassuto and Ryan | Oakland, 1970–75 | Agg Residential | TS+CS | L | −0.14 to −0.3 | |

**Table 2-5.** Price and Income Elasticities in North America (*Continued*)

| Date of publication | Authors | Study location | Type of use | Type of data | Type of model | Own-price elasticity | Income elasticity |
|---|---|---|---|---|---|---|---|
| 1979 | Danielson | 261 households in Raleigh, NC, 1969–74 | SF Residential, annual | CS+TS | LL | −0.27 | 0.33 |
| | | 261 households in Raleigh, NC, 1969–74 | SF Residential, winter | CS+TS | LL | −0.3 | 0.35 |
| | | 261 households in Raleigh, NC, 1969–74 | SF REsidential, summer | CS+TS | LL | −1.38 | 0.36 |
| 1979 | Foster and Beattie | 217 US cities | M&I | CS | SL | −0.53 | 0.18 |
| | | New England | M&I | CS | SL | −0.43 | |
| | | Midwest | M&I | CS | SL | −0.3 | |
| | | South | M&I | CS | SL | −0.38 | |
| | | Plains | M&I | CS | SL | −0.58 | |
| | | Southwest | M&I | CS | SL | −0.36 | |
| | | Pacific Northwest | M&I | CS | SL | −0.69 | |
| 1979 | Male et al. | Eastern United States | M&I | CS | L | −0.2 | 0.25 |
| | | Eastern United States | M&I | CS | LL | −0.68 | 0.46 |
| | | Eastern United States | M&I | CS | SL | −0.35 | 0.55 |
| 1980 | Agthe and Billings | Tucson, Arizona 1974–77 | Agg Residential | TS | L | −0.18, −0.36 | |
| | | Tucson, Arizona 1974–77 | Agg Residential | TS | LL | −0.18, −0.26 | |
| 1980 | Ben-Zvi | Red River Basin | Agg Residential, annual | CS | LL | −0.73 | |
| | | Red River Basin | Agg Residential, winter | CS | LL | −0.79 | |
| | | Red River Basin | Agg Residential, summer | CS | LL | −0.82 | |
| 1980 | Billings and Agthe | Tucson, Arizona 1974–77 | Agg Residential | TS | LL | −0.27 | |
| | | Tucson, Arizona 1974–77 | Agg Residential | TS | L | −0.49 | |
| 1980 | Carver and Boland | 13 areas | M&I, winter, long-run | TS+CS | L | −0.7 | |

| Year | Author | Sample | Category | Data | Functional form | Price elasticity | Income elasticity |
|---|---|---|---|---|---|---|---|
| 1980 | Morris and Jones | 13 areas | M&I, winter, short-run | TS+CS | L | −0.05 | |
| | | 13 areas | M&I, summer, long-run | TS+CS | L | −0.11 | |
| | | 13 areas | M&I, summer, short-run | TS+CS | L | −0.1 | |
| | | Households | SF Residential, annual | CS | LL | −0.39 | |
| | | Households | SF Residential, winter | CS | LL | −0.09 | |
| | | Households | SF Residential, summer | CS | LL | −0.73 | |
| 1981 | Foster and Beattie | United States | M&I | CS | SL | −0.47 | 0.46 |
| 1981 | Hansen and Narayanan | Salt Lake City, 1961–77 | M&I | TS | LL | −0.47 | |
| 1982 | Billings | Tucson, Arizona 1974–77 | M&I | TS | L | −0.66 | 2.14 |
| | | Tucson, Arizona 1974–77 | M&I | TS | LL | −0.56 | |
| 1982 | Howe | United States | SF Residential | CS | | −0.06 | |
| | | United States, east | SF Residential, summer | CS | | −0.57 | |
| | | United States, west | SF Residential, summer | CS | | −0.43 | |
| | | United States, west | | | | | |
| 1982 | Morgan | 473 households in Oxnard, CA | SF Residential | CS+TS, monthly | L | | 0.46 |
| 1984 | Jones and Morris | 326 households | SF Residential | CS | LL | −0.21 | |
| 1987 | Moncur Honolulu | 1281 households in | SF Residential | CS+TS, monthly | L | | 0.38 to 0.8 |
| 1989 | Billings and Day | 11 districts in Tucson, AZ, 1974–80 | Agg Residential | CS+TS | | | 0.36 |
| 1989 | Weber | 12 sub-districts of EBMUD, Oakland | Agg Residential, winter | TS+CS | L | −0.2 | |
| | | 12 sub-districts of EBMUD, Oakland | Agg Residential, winter | TS | L | −0.08 | |

**Table 2-5.** Price and Income Elasticities in North America (*Continued*)

| Date of publication | Authors | Study location | Type of use | Type of data | Type of model | Own-price elasticity | Income elasticity |
|---|---|---|---|---|---|---|---|
| 1990 | Boland, McPhail, and Opitz | Households in Southern California | SF Residential, winter | CS+TS | LL | −0.01 to −0.02 | |
| | | Households in Southern California | SF Residential, summer | CS+TS | LL | −0.13 to −0.18 | |
| 1991 | Dziegielewski and Opitz | Metropolitan Water District of So. Calif. | SF Residential, winter | TS+CS | LL | −0.24 | |
| | | Metropolitan Water District of So. Calif. | SF Residential, summer | TS+CS | LL | −0.39 | |
| | | Metropolitan Water District of So. Calif. | MF Residential, winter | TS+CS | LL | −0.13 | |
| | | Metropolitan Water District of So. Calif. | MF Residential, summer | TS+CS | LL | −0.15 | |

SOURCE: Based on Tables A1–A4 in Dziegielewski and Opitz (1991), with modifications.

KEY: SF = single-family; MF = multifamily; Agg. = aggregate; CS = cross-section; TS = time series; L = linear demand, (54); LL = log-log demand, (56); SL = semilog demand, (57).

Billings, R. B. and W. M. Day. 1989. Demand Management Factors in Residential Water Use: The Southern Arizona Experience. *Journal of the American Water Works Association,* 81(3): 58–64.

Boland, John J., Alexander A. McPhail, and Eva M. Opitz. August 1990. *Water Demand of Detached Single-Family Residences: Empirical Studies for the Metropolitan Water District of Southern California.* Carbondale, IL: Planning and Management Consultants, Ltd.

Brown and Caldwell. 1984. *Residential Water Conservation Projects:* Summary Report for U.S. Department of Housing and Urban Development. Walnut Creek, CA, June.

California Department of Water Resources. 1982. *Water Use by Manufacturing Industries in California, 1979.* Bulletin 124-3, May.

Camp, R. C. 1978. The Inelastic Demand for Residential Water: New Findings. *Journal of the American Water Works Association,* 70(8): 453–458.

Carver, P. H. 1978. *Price as a Water Utility Management Tool Under Stochastic Conditions.* Ph.D. Diss. Department of Geography and Environmental Engineering, Johns Hopkins University, Baltimore, MD.

Carver, P. H. and J. J. Boland. 1980. Short Run and Long Run Effects of Price on Municipal Water Use. *Water Resources Research,* 16(4): 609–616.

Cassuto, A. E. and S. Ryan. 1979. Effect of Price on Residential Demand for Water Within an Agency. *Water Resources Bulletin,* 15(2): 345–353.

Chesnutt, T. W., A. Bamezai, and C. N. McSpadden. 1992. *Continuous-Time Error Components Models of Residential Water Demand.* Report prepared for the Metropolitan Water District of Southern California. A&N Technical Services, Inc., Santa Monica, California.

Clark, Robert M. and Haynes C. Goddard. 1977. Cost and Quality of Water Supply. *Journal of the Water Works Association.* 69(1): 13–15.

Conley, B. C. 1967. Price Elasticity of Demand for Water in Southern California. *Annals of Regional Science.* I(1): 180–189.

Danielson, L. E. 1979. An Analysis of Residential Demand for Water Using Micro Time-Series Data. *Water Resources Research,* 15(4): 763–767.

Dziegielewski, B. 1988. Urban Nonresidential Water Use and Conservation, in Marvin Waterstone and R. John Burt, eds., *Proceedings of the Symposium on Water Use Data for Water Resources Management.* American Water Resources Association, August, pp. 371–380.

Dziegielewski, Benedykt and Eva Opitz. 1991. Municipal and Industrial Water use in the Metropolitan Water District Service Area, Interim Report No. 4. Prepared for the Metropolitan Water District of Southern California, Los Angeles. Planning and Management Consultants, Ltd., Carbondale, IL, June.

Flack, J. E. 1965. *Water Rights Transfers—An Engineering-Approach.* Ph.D. diss. Stanford University. Palo Alto, CA.

Foster, H. S. and B. R. Beattie. 1979. Urban Residential Demand for Water in the United States. *Land Economics,* 55(1): 43–58.

Foster, H. S. and B. R. Beattie. 1981. On the Specification of Price in Studies of Consumer Demand under Block Price Scheduling. *Land Economics,* 57(4): 624–629.

Fourt, L. 1958. *Forecasting the Urban Residential Demand for Water,* Agricultural Economics Seminar Paper, Department of Economics, University of Chicago, Chicago, IL.

Gallagher, D. R. and R. W. Robinson. 1977. *Influence of Metering, Pricing Policy, and Incentives on Water Use Efficiency.* Technical Paper No. 19. Australian Water Resources Council. Canberra, Australia.

Gardner, B. D. and S. H. Schick. 1964. *Factors Affecting Consumption of Urban Household Water in Northern Utah.* Agricultural Experiment Station Bulletin. No. 449. Utah State University: Logan.

Gardner, R. L. 1977. An Analysis of Residential Water Demand and Water Rates in Minnesota. *Water Resources Research Center.* Bulletin 96. University of Minnesota: Minneapolis.

Gibbs, K. C. 1978. Price Variable in Residential Water Demand Models. *Water Resources Research.* 14(1): 15–18.

Gottlieb, M. 1963. Urban Domestic Demand for Water: A Kansas Case Study. *Land Economics,* 39(2): 204–210.

Griffin, Ronald C. and Chan Chang. 1991. Seasonality in Community Water Demand. *Western Journal of Agricultural Economics,* 16(2): 207–217.

Grima, A. P. 1972. *Residential Water Demand: Alternative Choices for Management.* Toronto, Canada: University of Toronto Press.

Grunwald, Orlen C., C. T. Haan, D. L. Debertin, and D. I. Carey. 1976. Rural Residential Water Demand in Kentucky: A Econometric and Simulation Analysis. *Water Resources Bulletin,* 12(5): 951–961.

Hansen, R. D., and R. Narayanan. 1981. A Monthly Time Series Model of Municipal Water Demand. *Water Resources Bulletin,* 17(4): 578–585.

Headley, J. C. 1963. The Relation of Family Income and Use of Water for Residential and Commercial Purposes in the San Francisco-Oakland Metropolitan Area. *Land Economics,* 39(4): 441–449.

Hittman Associates, Inc. 1970. *Price, Demand, Cost and Revenue in Urban Water Utilities.* Columbia, MD. NTIS PB 195 929.

Hogarty, T. F. and R. J. Mackey. 1975. The Impact of Large Temporary Rate Changes on Residential Water Use. *Water Resources Research,* 11(6): 791–794.

Howe, C. W. and F. P. Linaweaver. 1967. The Impact of Price on Residential Water Demand and Its Relation to System Design and Price Structure. *Water Resources Research,* 3(1): 13–32.

Howe, Charles W. 1982. The Impact of Price on Residential Water Demand: Some New Insights. *Water Resources Research,* 18(4): 713–716.

Jones, C. Vaughn and J. R. Morris. 1984. Instrumental Price Estimates and Residential Water Demand. *Water Resources Research,* 20(2): 197–202.

Katzman, M. J. 1977. Income and Price Elasticities of Demand for Water in Developing Countries, *Water Resources Bulletin,* 13(1): 47–55.

Kiefer, Jack C. and Benedykt Dziegielewski. 1991. *Analysis of Residential Landscape Irrigation in Southern California.* A Report prepared for the Metropolitan Water District of Southern California, Los Angeles. December, prepared by Planning and Management Consultants, Ltd., Carbondale, IL.

Larson, B. O. and H. E. Hudson Jr. 1951. Residential Water Use and Family Income. *Journal of the American Water Works Association,* 43(7): 605–611.

Maidment, David R. and Shaw-Pin Miaou. 1986. Daily Water Use in Nine Cities. *Water Resources Research,* 22(6): 845–885.

Male, J. W., C. E. Willis, F. J. Babin, and C. J. Shillito. 1979. *Analysis of the Water Rate Structure as a Management Option for Water Conservation.* Publication 112, Water Resources Research Center, University of Massachusetts, Amherst.

Metcalf, L. 1926. Effect of Water Rates and Growth in Population Upon Per Capita Consumption. *Journal of the American Water Works Association,* 15(1): 1–20.

Metropolitan Water District, *Urban Water Use Characteristics in the Metropolitan Water District of Southern California,* Planning Division, April 1993.

Moncur, J. T. 1987. Urban Water Pricing and Drought Management. *Water Resources Research,* 23(3): 393–398.

Morgan, W. D. 1974. A Time Series Demand for Water Using Micro Data and Binary Variables. *Water Resources Bulletin,* 10(4): 697–702.

Morgan, W. D. 1982. Water Conservation Kits: A Time Series Analysis of Conservation Policy. *Water Resources Bulletin,* 18(6): 1039–1042.

Morgan, W. D. and J. C. Smolen. 1976. Climatic Indicators in the Estimation of Municipal Water Demand. *Water Resources Bulletin,* 12(3): 511–518.

Morris, J. R. and C. V. Jones. 1980. *Water for Denver: An Analysis of the Alternatives.* Environmental Defense Fund, Inc. Denver, CO.

Pope, R. M., Jr., J. M. Stepp, and J. S. Lytle. 1975. *Effects of Price Change Upon the Domestic Use of Water Over Time.* Water Resources Research Institute. Clemson University, South Carolina.

Primeaux, W. J. and K. W. Hollman. 1973. Price and Other Selected Economic and Socio-Economic Factors as Determinants of Household Water Consumption. In *Water for the Human Environment: Proceedings of the First World Congress on Water Resources.* vol. 3. Champaign, IL: International Water Resources Association.

Seidel, H. F. and E. R. Baumann. 1957. A Statistical Analysis of Water Works Data for 1955. *Journal of the American Water Works Association,* 42(12): 1531–1566.

Sewell, W. R. Derrick and L. Roueche. 1974. Peak Load Pricing and Urban Water Management: Victoria, B.C., A Case Study. *Natural Resources Journal,* 14(3): 383–400.

Smith, James A. 1988. A Model of Daily Municipal Water Use for Short-Term Forecasting. *Water Resources Research,* 24(2): 201–206.

Turnovsky, S. J. 1969. The Demand for Water: Some Empirical Evidence of Consumers' Response to a Commodity Uncertain in Supply. *Water Resources Research,* 5(2): 350–361.

Varian, Hal R. 1992. *Microeconomic Analysis.* 3d ed. New York: Norton.

Varian, Hal R. 1996. *Intermediate Microeconomics: A Modern Approach.* 4th ed. New York: Norton.

Ware, J. E. and R. M. North. 1967. *The Price and Consumption of Water for Residential Use in Georgia. Bureau of Business and Economic Research.* School of Business Administration, Georgia State University, Atlanta, GA.

Weber, J. A. 1989. Forecasting Demand and Measuring Price Elasticity. *Journal of the American Water Works Association,* 81(5): 57–65.

Wong, S. T. 1972. A Model of Municipal Water Demand: A Case Study of Northeastern Illinois. *Land Economics,* 48(1): 34–44.

Young, R. A. 1973. Price Elasticity of Demand for Municipal Water: A Case Study of Tucson, Arizona. *Water Resources Research,* 9(4): 1068–1072.

water use, but is based on assumptions which also turn out to be incorrect, the method cannot be said to be inaccurate. Similarly, accuracy cannot be claimed for a method which produces a good estimate of water use, but is based on incorrect assumptions.

Even if accuracy could somehow be assessed, it would be incomplete as a criterion for selecting a forecasting method. Most forecasts, including water use forecasts, are themselves assumptions in a larger decision-making process. Some water use forecasts underlie long-range supply plans, others determine short-range operating plans, still others lead to revenue estimates. No forecast would be used for any of these purposes if it were not credible and persuasive. The planner or decision maker must believe that the forecast represents a likely future, otherwise no decision would be based on the forecast. The paradox here is that a complex but highly accurate method may be avoided, while a simplistic but unreliable method is quickly adopted by decision makers. This dilemma reflects the difficulty that many people have in accepting results from a "black box."

A further complication appears when decision makers express a need for "conservative" forecasts. The term *conservative* can mean many things, but in forecasting it often describes a prediction of future activity which is purposely different from the most likely state of affairs. "Conservative" water use forecasts are usually higher than the most likely value, so as to reduce the probability of constructing facilities which turn out to be inadequate. Planning in this way has a cost, of course. It leads to systematic overbuilding, and permanently higher capital costs. A forecast that was purposely skewed downward would also imply a cost: facilities would be systematically underbuilt, sacrificing economies of scale and increasing the probability of drought costs and rationing. In general, "conservative" forecasts are a bad idea. It is far better to predict the most likely future level of water use, with appropriate attention to the uncertainty implicit in that forecast (see discussion of uncertainty following).

## Correlates of Accuracy in Water Use Forecasts

It can be seen that accuracy is a desirable characteristic of a forecast method, but is not a practical criterion for choosing among methods. Furthermore, in some situations the appearance of accuracy (credibility, persuasiveness) may be more important than accuracy itself. Research has shown that, in the absence of direct observations of accuracy, forecast methods are evaluated on the basis of correlates of accuracy (Ascher 1978). What these correlates are considered to be depends on who is making the determination.

Some forecast methods are judged by "outsiders," decision makers or others who are the users of the forecast, but are not the peers of the analyst preparing the forecast. Outsiders do not generally possess the skills or training to prepare the forecast, but they must decide whether to use it or ignore it. Not surprisingly, outsiders tend to rely on prior knowledge of the general experience with the forecast method being proposed, or with the analyst. Outsiders' judgment is also influenced by the institutional base of the analyst (What organization does the analyst work for? What are the perceived biases of that organization?) as well as the professional training and experience of the analyst (modeler? economist? engineer?). It can be seen that the actual characteristics of the forecast method have little to do with this evaluation.

A much more explicit evaluation is likely to be performed by "insiders"—those peers of the analyst who can understand and compare various models, assumptions, and computations. Insider evaluations are based on a number of correlates of accuracy, such as:

- Proper choice of scope (Is the right variable being forecast? for an appropriate geographic area?)
- Adequate disaggregation (If the forecast variable can be broken down into components, has this been done?)
- Model and assumptions reflect the center of opinion (Is there general support for assumptions? Is the explanatory model generally accepted?)
- Minimum feasible use of subjective analysis (Does the method make effective use of all reasonably available data? Is analysis used wherever possible, instead of assumptions?)
- Appropriate compromise between cost and complexity (Is the method needlessly simplistic? Are available data overlooked?)
- Robust model and assumptions (Is the form of the model likely to be invalidated in the future? Are assumptions sensitive to unknown trends or conditions?)

Careful consideration of these topics can identify strong and weak points for most forecast methods, leading to a better informed choice of method in each situation.

## Evaluation of Water Use Forecast Methods

As noted earlier, forecast methods are distinguished by the way in which they explain past water use. In terms of complexity, this explanation can range from a pencil line drawn on graph paper to the most sophisticated econometric model. The general approaches described here are meant to capture the range of possible methods suggested by current knowledge. Some of these methods are widely used, some rarely applied, and still others are not yet fully developed.

### Time Extrapolation

Water use can be represented as a time series, with past observations fitted to a smooth curve by graphical or mathematical curve-fitting methods. Once the curve is fitted, forecasting is done simply by extending the curve into the future. Time extrapolation has usually been applied to aggregate water use but it could, in principle, be used for water use within a specific sector. However it is applied, the underlying assumption is that the level of water use is explained by the passage of time, and that all other variables (such as population, price, employment, etc.) are either uncorrelated with water use or perfectly correlated with time.

This method enjoyed some popularity early in the twentieth century, but is rarely used today. Review of some of the correlates of accuracy noted previously will suggest the reasons. The approach is highly subjective, using very little data. It attempts to explain water use in terms of a variable, time, which by itself explains nothing. The implicit assumptions regarding explanatory variables other than time are demonstrably incorrect. The method is much too simplistic for virtually any application, and is highly sensitive to changes in the structure of water use.

### Bivariate Models

It has been common practice to explain water use in terms of a single variable, usually population. This implies a simple bivariate model of water use which can be written, in linear form, as:

$$Q = a + b \cdot X \tag{1}$$

where: $Q$ = water use per unit time
$X$ = explanatory variable
$a, b$ = coefficients

This model can be applied to aggregate or disaggregate water use. That is, $Q$ may represent total urban water use, or it may be residential water use, or even water use within a specific user category. $Q$ may also be a measure of average annual water use, summer-season water use, or maximum daily water use. Most applications of this approach assume that $a = 0$.

**Per Capita Requirements Method.**   One possible variant of the basic bivariate model is the per capita requirements model, as follows:

$$Q = b \cdot P \tag{2}$$

where:   $Q$ = average daily aggregate water use
  $P$ = resident population in service area
  $b$ = per capita water use

Over the past 100 years, this method has been the most widely used explanation of urban water use. While it could be applied to disaggregate water use (residential water use, for example), the method is almost always used to explain aggregate use for an urban area.

As is the case for other bivariate explanations of water use, the per capita method assumes that a single explanatory variable (population) provides an adequate explanation of water use. Other variables are assumed to be either unimportant or perfectly correlated with population. Since these assumptions are implicit in the method, and are not generally tested, the method can be seen to have a large subjective content. The data requirements of this method are modest: only water use and population data are needed. Neither the model nor its coefficient can be described as robust. Per capita water use varies widely from place to place; it can be observed to rise and fall over time for any specific location. This kind of variation suggests that the per capita model is an inadequate explanation of urban water use. Forecasts based on this model have generally not been noted for their accuracy.

Other problems are associated with the customary application of this approach to aggregate water use. While residential use may be moderately well correlated with population (it is actually better correlated with number of housing units), the same cannot be said for commercial, institutional, industrial, or public uses of water. Each of these sectors responds to a particular set of variables, which may or may not include population. Many cities with rising residential water use have experienced falling industrial use, changes in commercial water use may reflect the role of a city as a regional center and have little to do with resident population, and so forth. The per capita coefficient conceals these trends and relationships, creating an inadequate basis for forecasting.

**Unit Use Coefficient Method.**   Notwithstanding the weaknesses of the per capita method, other applications of the bivariate model may prove useful in the context of disaggregate forecasts. As the sectors and categories that are to be forecast separately become smaller, it is more reasonable to explain water use in each category in terms of a single variable. For example, total employment may be a perfectly adequate explanation for water use in office buildings. The relationship between water use and employment can be specified as shown in equation (1), where $a$ may or may not be set to zero, and $b$ is described as a unit use coefficient.

Although there certainly are other variables which affect water use in these categories (price, weather, etc.), if the sector is small and/or data collection costs are high, the simple bivariate relationship may be the best compromise between cost and complexity. If the explanatory variable is chosen carefully, the coefficient may prove acceptably stable over time.

### Multivariate Models

Multivariate water use models can take the following form:

$$Q = a + b_1 \cdot X_1 + b_2 \cdot X_2 + \cdots + b_n \cdot X_n \tag{3}$$

where:  $Q$ = water use per unit time
$X_i$ = explanatory variable $i$
$a, b_1, b_2, \ldots b_n$, = coefficients

If the variables in equation (3) are replaced by their natural logarithms, then the log-linear model is the equivalent of the following multiplicative form:

$$Q = \alpha \cdot X_1^\beta \cdot X_2^\gamma \cdot X_n^\delta \cdots \tag{4}$$

where:  $\alpha, \beta, \gamma, \delta$ = coefficients

Where several variables affect the same kind of water use, a model in the form of equation (4) may be more appropriate. For example, if both number of housing units and price influence residential water use, their relationship is multiplicative. However, if number of housing units is thought to determine indoor water use, and price is expected to be the major variable explaining outdoor water use, the relationship is additive, as shown by equation (3). More complex specifications are possible, combining elements of both the additive and multiplicative forms.

**Multivariate Requirements Models.**   In economics, the term "requirements" implies that the demand for a good will be assumed independent of its price. In water-demand modeling, this term is often applied to model specifications that reflect observed correlation, rather than theory. For example, an analyst may collect data for a number of variables, then develop a model which exhibits the best fit to the available data, perhaps after deleting variables which do not appear well correlated with water use. Economic variables, such as price and income, may not be included. An example of a requirements model is the Installation Water Resources Analysis and Planning System (IWRAPS©) for military installations, which is discussed in Chapter 4 (note that on military installations, the price of water is not a relevant factor).

The use of a multivariate model reduces the degree of subjectivity in the analysis, and makes better use of available data. For these reasons, multivariate models generally represent an improvement on bivariate or simpler methods. The use of such models is consistent with disaggregate forecasting. For example, multivariate requirements models have been successfully applied to water use within categories of the manufacturing sector. Typical explanatory variables in this case include number of employees, value of output, and water recycle ratio.

The major disadvantage of this approach is the fact that the model reflects correlation rather than causation. The initial list of variables does not imply any theory of water use, so it may well omit potentially important relationships. There is also a substantial risk of incorporating spurious correlations, which may appear in historical data but are not likely to be replicated in the future. Requirements models, then, are not generally accepted as reasonable descriptions of water use, and may not be robust over time.

**Econometric Demand Models.**   The obvious alternative to the requirements model is one which is based on theory, and which expresses in its form the most likely causal relationships between explanatory variables and water use. Using the principles and techniques of econometrics, models of water use can be developed that describe the true economic demand for water. Such models include one or more variables representing the price and/or tariff structure, and where appropriate, may also include a variable describing water users' income or ability to pay.

So far, econometric demand models are available mostly for residential water use at detached single-family dwellings. Such models have been highly successful, and are able to accurately predict water use under a wide range of circumstances. This should not be surprising, after review of the correlates of accuracy listed previously. Econometric demand models, properly applied, satisfy each item on the list. An example of an econometric demand model is shown in Chapter 4 (i.e., the residential sector models of IWR-MAIN).

Unfortunately, few models of this kind have been developed for multi-family residential buildings, or for nonresidential water uses. Further research is needed before this approach can be used to forecast urban water use in all sectors.

## Uncertainty in Forecasts

Forecast water use quantities are subject to uncertainty for at least three general reasons:

- *Model misspecification*—The model may omit important explanatory variables, it may include unrelated (spurious) variables, or it may misrepresent the functional relationships existing between explanatory and dependent variables.

- *Coefficient error*—Model coefficients may be incorrectly estimated because of errors or other disturbances in past data, errors in procedure, or collinearity among included variables.

- *Assumption error*—Assumptions made at any point, including those pertaining to future values of explanatory variables, may be in error.

Forecasts should minimize risk and uncertainty. This can be done by choosing methods which reduce error due to model misspecification, and which make maximum use of objective analysis, thus reducing assumption error. Both of these steps, however, increase complexity and the number of explanatory variables, thus increasing exposure to coefficient error.

Those who prepare and use forecasts have long recognized the need to make allowance for uncertainty. Various approaches have been taken, although the most common methods are simplistic, demonstrating little awareness of the nature of uncertainty. The following paragraphs provide brief descriptions of some techniques for expressing risk and uncertainty in forecast results.

### "Safety Factors"

One way to express the uncertainty in a forecast is to arbitrarily add or subtract some fixed amount from the forecast value. A water use forecast may be increased by 10 percent, for example, to provide a "safety factor" or "margin for error." Application of this method requires the analyst to decide in which direction to adjust the forecast (up or down) and by how much. Forecasts of this kind are often spoken of as "conservative" forecasts.

The safety-factor method rests on the assumption that all errors of practical concern are on one side of the originally forecast value. If overestimates are feared, the forecast can be reduced by the safety factor; if only underestimates are to be avoided, the forecast is increased. Unfortunately, such a simplification is rarely, if ever, valid. Forecast errors of both types are potential problems.

Where the purpose of the forecast is to plan facilities to meet future demands, an underforecast will lead to early facility obsolescence and/or failure to meet some or all demands. In the first case, additional facilities must be built before the end of the planning period, perhaps uneconomically; in the second situation, lost consumer benefits and, perhaps, rationing costs must be faced. In either case, forecast error will result in excess cost. An overforecast also imposes excess cost. Building facilities which are too large or built too soon increases the cost of a project without compensating increases in benefits.

In either case, the larger the error, the larger the cost. It is not clear that the cost of a given underestimate is larger than the cost of a comparable overestimate; neither is the converse necessarily true. The expected cost of forecast error is unlikely to be minimized by an arbitrary adjustment in an arbitrary direction.

Another way of inserting safety factors into a forecast, or of making a forecast conservative, is to make the necessary adjustments to the assumptions and coefficients, rather than the forecast results. Forecasts obtained directly from population forecasts that have been inflated, or using coefficients that are known to be too high, are also conservative forecasts, in the same way that upward revisions to the forecast value may be termed conservative.

To understand the meaning of "conservative" when applied to forecasts of this kind, it is necessary to identify the value that is being conserved. In the case of water use forecasts, where conventional practice defines a conservative forecast as one that is higher than it would be in the absence of "conservatism," the meaning is easy to discern. Priority has been given to protection against underestimates, indicating that the planner is primarily concerned with protecting his/her reputation for planning and building facilities that prove to be large enough. No concern for minimizing the expected cost borne by the public is evident.

## Scenario Approaches

Another way to express forecast uncertainty involves the preparation of a number of forecasts, each based on a different set of assumptions. A particular set of assumptions, or scenario, should be internally consistent, describing one of a number of possible states. In most applications, a relatively small number of scenarios are used (e.g., fewer than ten).

The scenario method is useful in conveying some sense of the range of possible outcomes. However, it can easily understate that range (plausible scenarios leading to more extreme forecasts can easily be overlooked). Also, no information is provided on the probability of any scenario actually occurring. Providing results for a number of unlikely scenarios, while omitting some more likely combinations of assumptions, may be misleading. Finally, the scenario approach is usually used to provide information on uncertainty due to assumption error only: no attention is ordinarily given to the possibility of model misspecification or coefficient error.

## Sensitivity Analysis

Rather than replacing one entire set of assumptions with another, as in the case of the scenario method, it is possible to examine individual assumptions one at a time. In this approach, known as sensitivity analysis, a selected assumption is varied, holding all others constant, while noting the effect on resulting forecast values. Repeating this process for most assumptions provides two kinds of information: (1) it identifies the critical assumptions; and (2) it indicates the range of possible forecast error.

While the general approach of sensitivity analysis can also be used for variations in model specification or coefficient estimates, it is most commonly applied to assumptions. In this form, sensitivity analysis, like scenario analysis, considers only one source of uncertainty. When fully exploited, sensitivity analysis is likely to yield a broader range of possible outcomes than scenario analysis, but it is similarly silent on the relative likelihood of any particular result. Further, sensitivity analysis is more likely to incorporate sets of assumptions which are mutually inconsistent, or at least highly unlikely (e.g., rising income coupled with falling employment).

## Contingency Trees

A further extension of the ideas contained in scenario and sensitivity analyses leads to the use of contingency trees. In this case, a number of key assumptions are identified and each is restated as two or more alternate assumptions. A subjective probability of occurrence is then assigned to each alternate assumption. A contingency tree is constructed showing all possible combinations of alternate values for the assumptions under study. The joint (subjective) probability of each combination of alternatives is determined, as is the associated forecast outcome. This information can be used to present a probability distribution of possible outcomes.

The results of this method, of course, reflect only the uncertainty contributed by the assumptions under study, and do so only to the extent that the analyst is able to supply estimates of probabilities. Uncertainty due to model misspecification or coefficient error are not easily addressed in this way. Probabilities are difficult to estimate, especially when the assumptions are not independent. Still, the result is substantially more information than obtained from simpler methods.

## Forecast Bounds

**Arbitrary Bounds.**   Many forecasters attempt to provide upper and lower bounds for future values. In some cases, this is done by selecting a set of assumptions that will lead to a value lower than that considered likely, and another set that will produce a higher value. When these alternative assumptions are chosen to delineate the maximum range of plausibility, the resulting forecasts may provide realistic bounds. However, bounds chosen in this way may define a very wide interval which provides little useful information.

Other approaches to choosing forecast bounds may reflect criteria similar to those used to set safety factors, except that errors in both directions are considered. In either case, the bounds are arbitrary and convey no information about the likelihood of a result in the vicinity of either bound.

**Confidence Intervals.**   Another approach to the setting of bounds is the use of past data to develop objective estimates of the probability of errors of various sizes. This permits the definition of bounds which enclose, for example, the 95 percent confidence interval. Together with a "most likely" forecast, these bounds convey considerable information to the forecast user.

In the case of regression models, the needed probability information may come from the distribution of error around the estimated regression line. For some kinds of time-series models, such as autoregressive-moving average (ARMA) models, random terms are included, which cause values estimated by the model to reproduce distribution parameters estimated from the past data.

Properly applied, probabilistic methods are capable of reflecting uncertainty due to error from coefficient error and, where sufficient data are available, assumption error. The consequences of model misspecification are more difficult to represent, and are not usually amenable to statistical analysis of this type.

# Forecasting Under Uncertainty

## Base Forecasts

Consideration of risk and uncertainty in forecasting requires a base forecast, which serves as a reference point for various estimates of error. To be useful in this role, the base forecast should represent the most likely future value of the subject activity. Accordingly, it should make use of the forecasting method and assumptions considered most likely to yield an accurate result, avoiding all known sources of bias. A base forecast, in other words, should contain no safety factors or "conservative" assumptions.

Accuracy, as noted earlier, is associated with a number of forecast characteristics, including:

- Proper choice of scope
- Adequate degree of disaggregation, where feasible
- Model and assumptions which reflect center of opinion
- Minimum feasible use of subjective analysis
- Appropriate trade-off between complexity and cost
- Stable (robust) model and assumptions

Unless the base forecast is the most accurate possible in the circumstances, measures of risk and uncertainty will be unnecessarily large and difficult to interpret.

## Expressing Uncertainty

The degree to which risk and uncertainty can be identified and measured depends on the characteristics of the base forecast method and on additional information available to the analyst. In general, the more detailed the method, the more explicit the assumptions. More explicit assumptions permit more systematic examination of sources and degree of uncertainty. Simplistic base forecasts, on the other hand, permit only the most general statements to be made about uncertainty.

Approaches that omit probabilistic information (e.g., safety factors, scenarios, sensitivity analysis, arbitrary bounds), while superficially attractive because of minimal data requirements, should be used with caution. In most cases, there is substantial danger of misleading forecast users (by failing to explore an adequate range of possibilities) or of producing a range of outcomes so wide as to be meaningless. Even subjective probability estimates, if carefully developed, are preferable to none.

## Interpretation of Results

None of the approaches described here is capable of reflecting all of the risk and uncertainty that is normally present in forecasts. Most methods can reveal a portion of the uncertainty inherent in the forecast assumptions or in the model coefficients; some methods can address both subjects. Little can be said about the uncertainty resulting from model misspecification, and nothing is known of the possibility that the chosen model will no longer be appropriate at some future time.

The usefulness of forecasts in planning and evaluation is greatly improved when information on risk and uncertainty can be provided.

This information should be developed with respect to the most reliable possible base forecast, and should be probabilistic in nature wherever feasible. At the same time, it is important that the limitations of such information be made clear.

## References

Ascher, W. 1978. *Forecasting Methods: An Appraisal for Policy-Makers and Planners.* Baltimore, MD: Johns Hopkins University Press.

Boland, J. J., R. W. Wentworth, and R. C. Steiner. 1982. Forecasting short-term revenues for water and sewer utilities. *Journal American Water Works Association,* 74(9): 460–465.

Dziegielewski, B. and J. J. Boland. 1989. Forecasting urban water use: the IWR-MAIN model. *Water Resources Bulletin,* 25(1): 101–119.

Encel, S., P. K. Marstrand, and W. Page. 1975. *The Art of Anticipation: Values and Methods in Forecasting.* New York: Pica Press.

Jones, C. V., J. J. Boland, J. E. Crews, C. F. DeKay, and J. R. Morris. 1984. *Municipal Water Demand: Statistical and Management Issues.* Boulder, CO: Westview Press.

# 4

# Forecasting Urban Water Use: Models and Application

Eva M. Opitz, John F. Langowski,
Benedykt Dziegielewski, Nancy A.
Hanna-Somers, J. Scott Willett,
and Richard J. Hauer

*Planning and Management
Consultants, Ltd.*

This chapter addresses selected approaches and tools for forecasting urban water use. First, the IWR-MAIN Water Demand Analysis Software is a tool for estimating future water demands and evaluating water demand management measures in urban areas. Second, an end-use approach for an analysis of water demands and for forecasting is further elaborated. Third, a tool for forecasting water demands on military installations is presented (i.e., the Installation Water Resources Analysis and Planning System).

## IWR-MAIN Water Demand Analysis Software

### Background

The IWR-MAIN System is a computer software program for the preparation of water use forecasts and the analysis of demand management alternatives. The original version of the IWR-MAIN System was developed in 1969 by Hittman Associates, Inc., of Columbia, Maryland, under the sponsorship of the Office of Water Resources (later Office of Water Research and Technology) of the U.S. Department of Interior. Released as MAIN II, the program was based upon the early findings of a comprehensive water use study conducted at the Johns Hopkins University. This study led to the development of a set of water use models published by Howe and Linaweaver (1967). In 1982, the Institute for Water Resources (IWR) of the U.S. Army Corps of Engineers selected MAIN II as a tool for improving water use forecasting procedures within the Corps. Under sponsorship of IWR, a substantial research effort was undertaken in order to update the MAIN II program and modify it for easy access on personal computers. The product of these efforts was a software package, released in 1987, for personal computers called the IWR-MAIN Water Use Forecasting System, Version 5.1. The acronym IWR-MAIN stands for Institute for Water Resources—Municipal And Industrial Needs.

In conjunction with water supply studies, some Corps Districts have applied IWR-MAIN Version 5.1 to specific municipal areas. IWR-MAIN Version 5.1 (and variants thereof) have also been used by major water utilities in both the eastern and western United States including, among others:

1. Indianapolis Water Company (27 subdistricts)

2. Phoenix Water Services Department (4 study areas)

3. Metropolitan Water District of Southern California (57 study areas)

4. El Paso Water Utility

5. Binghamton, New York

6. Springfield City Water, Light, and Power (3 study areas)

7. Southwest Florida Water Management District (62 study areas)

8. Las Vegas Valley Water District (6 utilities)

9. City of San Diego Water Utilities Department

In 1994, under the sponsorship of the Institute for Water Resources, the Phoenix Water Services Department, and the Metropolitan Water District of Southern California, the IWR-MAIN System has been upgraded to the IWR-MAIN Water Demand Analysis Software, Version 6©. This upgrade includes a benefit-cost module for analyzing water demand management measures. The new version also includes an expanded and up-to-date knowledge base on residential water use, nonresidential water use, and end-use parameters. An advanced user-interface provides users with a tool for the handling of data files, the batch execution of forecast runs, and the processing of forecast results. Version 6 of the software has been used in water-demand and conservation studies for the Southern Nevada Water Authority, the Massachusetts Water Resources Authority, and the Lee County Regional Water Supply Authority (Florida).

The IWR-MAIN Water Demand Analysis Software (IWR-MAIN Software) is a modern software package designed for: (1) translating demographic, housing, and business statistics (for cities, counties, or service areas) into estimates of existing water demands and (2) using projections of population, housing, and employment to derive baseline forecasts of water use. The forecast module disaggregates total urban water use into spatial, temporal, and sectoral components (see Table 4-1). The water demands of each sector in a given area and time period are expressed as a product of (1) the number of users (i.e., demand drivers such as the number of residents, housing units, employees, parkways, etc.), and (2) the average rate of water use (e.g., per household or per employee) as determined by a set of explanatory variables, which, in the residential sector, may include income, price, household size, housing density, air temperature, and rainfall. The econometric equations for estimating average rates of water use conform to the economic theory of demand. The following summarize the major features of the IWR-MAIN Software:

1. *Spatial disaggregation:* The definitions of the study areas for which water use forecasts can be prepared are limited only by the availability of demographic input data for the defined geographic study areas. Forecasts can be prepared for drainage basins, counties, water utility service areas, census tracts, and other geographic configurations. Spatially

**Table 4-1.** IWR-MAIN Disaggregation of Urban Water Use

| Spatial | Temporal | Sectoral |
|---|---|---|
| Geographic area | Average-daily | Residential |
| Counties | Winter season |   Single-family |
| Watershed basins | Summer season |   Multifamily low-density |
| Water utility service areas | Maximum-daily |   Multifamily high-density |
| | |   Etc. |
| | | Commercial |
| | |   Retail business |
| | |   Service establishments |
| | |   Etc. |
| | | Manufacturing |
| | | Governmental |
| | | Public/user-added |
| | | Other/unaccounted |

disaggregating water use forecasts allow planners to account for varying growth rates among study areas.

2. *Seasonal disaggregation:* Forecasts of water use can be prepared for annual average-daily demands, summer average-daily demands, winter average-daily demands, and maximum-daily demands. This allows users to account for seasonal and maximum-daily variations of water use for facility and demand management planning.

3. *Sector disaggregation:* Forecasts of water use can be prepared for major sectors of water users including residential, nonresidential (manufacturing, commercial, and governmental), public, and other. If desired, users of the IWR-MAIN Software may further disaggregate water use into categories of use such as single-family homes, multifamily homes, hotels/motels, schools, bottling plants, and so forth. The user has the option of selecting the level of sector disaggregation.

4. *Determinants of water demand:* Estimates of water use account for the various factors that affect the rate of water use (per household, per employee, etc.). In the residential sector, these factors may include household income, household size, housing density, weather conditions, and the price of water and wastewater services. In the nonresidential sector, these factors may include employment by industry type, labor productivity, weather conditions, and the price of water and wastewater services.

5. *User-added categories:* The software allows the development of user-added categories of water use and the incorporation of their unit-use coefficients into IWR-MAIN. The user-added categories may include water uses that are either not explicitly addressed by the current water use models (e.g., water withdrawals for agricultural irrigation or mining) or water uses that the user wishes to address separately (e.g., public parks, public golf courses, etc.).

6. *Demand management:* The software has the ability to address the long-term water savings impacts of different demand management (conservation) practices. These water savings can be incorporated into long-term forecasts of water demand. Furthermore, a new module of the IWR-MAIN Software facilitates the conduct of benefit/cost analyses of demand management alternatives.

7. *Sensitivity analyses:* The software provides the ability to conduct numerous "what-if" scenarios regarding projected changes in the determinants of water demand and to assess their impact on long-term water demands.

The following sections describe the water use forecasting and conservation savings methods of the various sectors of current version of the IWR-MAIN Water Demand Analysis Software.

## Forecasting Methods

**Residential Sector.**   Within the residential sector of IWR-MAIN, there are seven subsectors available for forecasting water demands. These include:

1. Single-family—1 attached, 1 detached units

2. Multifamily low-density—2, 3, 4 units per structure

3. Multifamily high-density—5 or more units per structure

4. Mobile homes

5. Nonurban

6. User-added

7. Total residential

These categories correspond to the housing types used by the U.S. Bureau of the Census.

Average rates of water use within each residential subsector are estimated using causal water demand models which take the following form:

$$q_{c,s,t} = \alpha\, I^{\beta_1}\, H^{\beta_2}\, L^{\beta_3}\, T^{\beta_4}\, R^{\beta_5}\, P^{\beta_6}\, e^{b_7 B} \tag{1}$$

where   $q_{c,s,t}$ = predicted average water use in sector $c$, during season $s$, in
year $t$
$I$ = median household income
$H$ = average household size (persons)
$L$ = average housing density (units per acre)
$T$ = average maximum-daily temperature
$R$ = rainfall
$P$ = marginal price of water (including sewer charges related to water use)
$B$ = fixed charge or rate premium
$\alpha$ = constant
$\beta_i$ = elasticities of explanatory variables
$b_7$ = coefficient of the rate premium
$e$ = base of the natural logarithm

For the purpose of clarity, the seasonal, sectoral, and time indices of the explanatory variables in the preceding equation are suppressed.

The default elasticities for the explanatory variables of residential household water use are derived from econometric studies of water demand through a meta-analysis of empirical literature.* Approximately 60 studies of residential water demand, which contained almost 200 empirically estimated water use equations, were integrated in order to derive a set of unbiased estimates of long-term elasticities for each explanatory variable. These unbiased estimates are derived using a rigorous statistical technique (meta-analysis). This technique employs multiple regression to explain the variance in the values of the reported elasticities due to interstudy differences. Once the elasticities were selected, the intercept term was estimated by regressing the product of the explanatory variables (raised to their respective elasticities) on observed water use in each residential category from a sample of utilities around the United States. Table 4-2 shows the default IWR-MAIN elasticities for select residential subsectors.

**Nonresidential Sector.** The nonresidential sector of IWR-MAIN addresses water uses within the following major industry groups:

1. Construction

2. Manufacturing

* An elasticity is a dimensionless measure of the relationship between quantity (water use) and any explanatory variable (e.g., price, income). The elasticity is interpreted as the percent change in quantity (e.g., water use) that is expected from a 1 percent change in the explanatory variable. For example, an elasticity of +0.4 on income in a water-demand equation indicates that a 1 percent increase in income will cause a 0.4 percent increase in water use.

**Table 4-2.** Coefficients and Elasticities of Residential Water Use Currently Contained in IWR-MAIN Version 6*

| Explanatory variable | Single-family summer | Single-family winter | Multifamily low-density summer | Multifamily low-density winter |
|---|---|---|---|---|
| Income | +0.40 | +0.40 | +0.40 | +0.40 |
| Persons per household | +0.40 | +0.45 | +0.40 | +0.45 |
| Housing density | −0.65 | −0.30 | −0.30 | −0.15 |
| Marginal price | −0.25 | −0.04 | −0.15 | −0.02 |
| Fixed charge | −0.0005 | −0.0005 | −0.0005 | −0.0005 |
| Maximum-daily temperature | +1.50 | +0.45 | +1.20 | +0.35 |
| Total rainfall | −0.25 | −0.02 | −0.10 | −0.01 |

SOURCE: Planning and Management Consultants, Ltd. (1996).

* The meta-analysis of econometric water demand studies is continuing with new additions to the data base.

3. Transportation, communications, and utilities (TCU)

4. Wholesale trade

5. Retail trade

6. Finance, insurance, and real estate (FIRE)

7. Services

8. Public administration

The eight major industry groups are classified according to the Department of Commerce Standard Industrial Classification (SIC) codes. The model of water use in the nonresidential (commercial/industrial) sector is:

$$Q_i = f(GED_i, E_i, L_i, P_i, CDD, O_i) \tag{2}$$

where:    $Q_i$ = category-wide water use in gallons per day
$GED_i$ = per-employee water use in gallons per employee per day
$E_i$ = category-wide employment
$L_i$ = average productivity (of labor) in category $i$
$P_i$ = marginal price of water and wastewater services in category $i$
$CDD$ = cooling degree days
$O_i$ = other variables known to affect commercial/industrial water use

Although this theoretical model is fully operational within IWR-MAIN, there are no currently available econometric (and generally applicable) models that contain model elasticities for price, productivity, cooling degree days, or other variables. The software was designed to accommodate the model specification once data were available regarding the responsiveness of nonresidential water use to such variables. Currently, the default calculation is based upon gallon per employee per day coefficients for SIC categories and groups:

$$Q_i = (GED_i * E_i) \tag{3}$$

The water-use-per-employee coefficients contained within IWR-MAIN are the result of 10 years of research effort devoted to collecting data on employment and water use for various establishments throughout the United States. The water use coefficients within IWR-MAIN are based upon the analysis of water use and employment relationships in over 7000 establishments. Table 4-3 shows the water-use-per-employee coefficients for the eight major industry groups.

Within each major industry group, SIC codes distinguish more homogeneous groups at the two-digit SIC level and even further at the three-

**Table 4-3.** Nonresidential Water Use Coefficients Currently
Contained in IWR-MAIN

| Major industry group | SIC codes | Water use coefficient (gallons/employee/day)* | |
|---|---|---|---|
| Construction | 15–17 | 20.7 | (244) |
| Manufacturing | 20–39 | 132.5 | (2784) |
| Transportation, communications, utilities (TCU) | 40–49 | 49.3 | (225) |
| Wholesale trade | 50–51 | 42.8 | (750) |
| Retail trade | 52–59 | 93.1 | (1041) |
| Finance, insurance, real estate (FIRE) | 60–67 | 70.8 | (233) |
| Services | 70–89 | 137.5 | (1870) |
| Public administration | 91–97 | 105.7 | (25) |

SOURCE: Planning and Management Consultants, Ltd. (1996).

* The numbers in parentheses represent the sample number of establishments from which the water use coefficient was calculated.

digit SIC level. For example, within the Services major industry group, there are 15 two-digit SIC categories which represent different types of establishments such as Hotels and Other Lodging Places (SIC 70) or Personal Services (SIC 72). Within each two-digit SIC level, the establishment types can be further disaggregated into even more homogeneous groups. For example, within Personal Services (SIC 72) are Laundry Facilities (SIC 721) or Beauty Shops (SIC 723).

In total, IWR-MAIN can address water uses in eight major industry groups, 65 two-digit SIC categories, or 417 three-digit SIC categories. Water use per employee coefficients are available for each category. For each major industry group, IWR-MAIN users have the option of addressing water uses at the major industry group level, two-digit SIC level, or three-digit SIC level. The decision of the appropriate level to be used for a particular study area should be based upon the structure of the employment in a community, the availability of employment data at the various levels, and the need for or desire to conduct sensitivity analysis regarding potential water use impacts.

**Public Sector.** The public sector within IWR-MAIN can be used to address water uses that occur in a service area which were not accounted for by the other sectors. These categories may address special-purpose water needs and are considered to be user-added categories (i.e., the IWR-MAIN Software user must specify all necessary parameters for the water use calculation). Water use in these user-added categories needs to be expressed as a function of a single explanatory variable (such as the number of acres or number of facilities). Water use coefficients (expressed in gallons per day per selected unit) must be provided by the IWR-MAIN Software user. Projections of water use can then be made by projecting the number of units into the future. The following is a list of uses and parameters for possible "public" categories:

- Irrigation of public parks and medians, acres
- Make-up water for public swimming pools, number of pools
- Irrigation of golf courses

**Other/Unaccounted Sector.**   The difference between total water production (i.e., total water into distribution system) and the quantity of water sold is typically referred to as unaccounted-for water use. This unaccounted-for sector may represent different types of use/loss including:

1. Distribution system leakage
2. Meter misregistration
3. Hydrant flushing
4. Major line breaks
5. Firefighting
6. Unmetered or nonbilled customers
7. Illegal connections
8. Street washing/construction water

Water utilities typically report unaccounted-for water use as a percentage difference between total water into the system and total metered sales. Within IWR-MAIN, users specify a percentage rate for each base and forecast year to estimate the amount of current and future unaccounted-for water.

### Conservation Savings Methods

In order to enhance the ability of water planners to formulate, implement, and evaluate various demand management alternatives, the observed sectoral demands during a defined season of use can be disaggregated into their applicable end uses. Only such a high level of disaggregation will permit water planners to make all necessary determinations in estimating water savings of various programs (Dziegielewski and Strus 1993; Dziegielewski et al. 1993).

The end uses represent the specific purposes of water use that are found in urban areas. For example, indoor residential end uses include toilet flushing, showering, bathing, bathroom faucet use, dishwashing, clothes washing, kitchen faucet use, and water use through other indoor outlets. The outdoor residential end uses include landscape irrigation, swimming pool use, Jacuzzi use, car washing, and other. Industrial and commercial sectors usually have some additional end uses such as water-cooled air conditioning, cooling and condensing, boiler use, and process use. The significant residential and nonresidential end uses have to be quantified in order to determine the potential for improving the efficiency of water use.

The conservation savings module of the IWR-MAIN Software uses the end-use accounting procedure to disaggregate seasonal demands of various water use sectors into a number of specific end uses such as dishwashing, toilet flushing, lawn watering, cooling, and others. This high level of disaggregation is designed to accommodate the evaluation of various demand management (conservation) measures which usually target specific end uses. The IWR-MAIN conservation savings module utilizes an end-use accounting system which disaggregates the seasonal demands of various water use sectors into as many as 20 different end uses. A rational representation of each end use is illustrated in Figure 4-1. Given parameters of local end-use conditions, this procedure predicts the average quantity of water in each end use as a function of: (1) the distribution of end uses among three classes of efficiency (i.e., nonconserving, standard, and ultraconserving), (2) average usage rate or intensity of use, (3) leakage rate and incidence of leaks, and (4) presence of end use within a given customer sector. The structure of the end-use model allows the planner to estimate the net effects of long-term conservation programs by tracking the values of each end-use parameter over time.

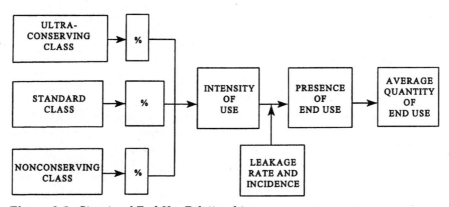

**Figure 4-1.** Structural End-Use Relationships

The end-use relationship may be conveniently represented by a single formula. End uses may be calculated as:

$$q_i = [(M_1S_1 + M_2S_2 + M_3S_3) \cdot U_N + K \cdot F_N] \cdot A_N \tag{4}$$

where: $q_i$ = quantity of water used by end use $i$, GPD/unit
$M_{1-3}$ = mechanical parameter (e.g., volume per use, flow rate per minute)
$S_{1-3}$ = fraction of the sector for end use that is nonconserving, standard, and ultraconserving
$U_N$ = intensity of use parameters (e.g., flushes per day/unit, minutes of use per day/unit)
$K$ = mechanical parameter representing the rate of leakage
$F_N$ = fraction of end uses with leakage
$A_N$ = fraction of units in which end use $i$ is present
$N$ = normal use or nondrought/nonemergency
1–3 = end use or group that is nonconserving, standard, and ultraconserving, 1 signifying the lowest level of efficiency

As designated in the equation, within each end use there are three classes of efficiency (e.g., inefficient, standard, and conserving). Each efficiency class is defined by the mechanical parameters ($M_1$, $M_2$, and $M_3$). The percent (i.e., fraction) of all end uses within a sector in each efficiency class is designated as $S_1$, $S_2$, and $S_3$, respectively.

The intensity of use for each purpose, $U$, defines how frequently a particular end use occurs or what is the average intensity of use. In the residential sector, an event or flow rate can be defined on a per person basis. For example, a toilet is flushed an average of four to five times per person per day. This value, when multiplied by the average number of persons per household in the study area, gives the intensity value for the toilet end use. The leakage rate, $K$, and the incidence of leaks, $F$, capture the amount of water that is lost with each end use because of small leaks. For example, 5 to 20 percent of toilets may have slow leaks of 0.5 to 1.0 gallons per hour. Finally, the presence of a particular end use, $A$, represents the fraction of units (i.e., water customers) in a sector that have that particular end use.

An application of Equation (4) to the toilet end use in the residential sector requires the knowledge of all these parameters. For example, the efficiency classes defined using the typical values for the design parameter of 5.5, 3.5, and 1.6 gallons per flush may be distributed with the corresponding fractions of end uses within each class at 0.35, 0.55, and 0.10. Also, with a typical value of five flushes per person per day and average household size of 2.8 persons the usage rate for the toilets would be 14 flushes per day per home. An average leakage rate of 20 gallons per day can be assumed with the incidence of leaks of 0.15. The presence parameter for toilets is 1.0. Using Equation (4), the average end use of water for toilet flushing would be 59.1 gallons per day per house (i.e., [(5.5 * 0.35 + 3.5 * 0.55 + 1.6 * 0.10) * 14 + (20 * 0.15)] * 1.0). Other end uses can be represented using similar parameters and data.

The structural end-use relationship (4) is dictated by the need to distinguish between changes in water demand caused by demand management programs for the changes caused by other factors. The distribution of end uses among efficiency classes, average intensity, incidence of leaks and presence will change in response to changes in the determinants of water use such as income, household size, and housing density. For example, changes in price will cause a decrease in the incidence of leaks in the short run and will affect the distribution of end uses among the classes of efficiency in the long run. The other two parameters of the end-use equation (i.e., intensity and presence) also will be affected by changes in price.

Long-term conservation savings would be achieved by increasing the fractions $S_2$ and $S_3$. This is accomplished by moving customers from one efficiency class to another. For example, for each end use, the fraction of the water users would be shifted from nonconserving to ultraconserving or from standard to ultraconserving. The quantifiable effect of the program is accounted for directly by the numerical shift in the customer pools and the change in the fractions of customers in each efficiency class.

The conservation savings module distinguishes among passive, active, and emergency (i.e., temporary) conservation effects. Passive conservation effects are represented by shifts in end-use consumption from less efficient fixtures to more efficient fixtures brought about primarily by plumbing codes for new construction (e.g., the toilet end use moves from the inefficient 5.5 gallon-per-flush toilet, to the standard 3.5 gallon toilet, or the highly efficient 1.6 gallon toilet). The conservation savings of active programs are estimated by noting the changes in the distribution of efficiency classes brought about by the participation in a utility-sponsored program whose inefficient or standard end uses were replaced or retrofitted.

### Benefit/Cost Analysis Methods

The implementation of water demand–management (or conservation) programs can yield monetary savings (or economic benefits) both to the utility and to water customers. The IWR-MAIN benefit-cost procedure provides a screening mechanism for choosing economically viable program alternatives. Various conservation alternatives can be compared in terms of their costs and benefits. Furthermore, the results of the benefit-cost procedure can be used in comparing supply augmentation alternatives (e.g., the construction of a new reservoir) with demand management alternatives using the same economic criteria. The benefit-cost procedure is conceptually very simple, but its application can become quite involved because of the many benefit-cost items, different accounting perspectives, and economic feasibility tests that can be considered.

IWR-MAIN contains 26 possible benefit-cost items which are broken down into three groups: (1) deferred capacity expansion costs, (2) avoided variable costs, and (3) other program benefits and costs. These benefit-cost items may represent either benefits or costs, depending on the accounting perspective. Each study area and each conservation program to be evaluated will have unique characteristics that will determine which benefit-cost items will be included in the analysis.

The economic tests associated with the various conservation programs will vary depending on the accounting perspective under which the analysis is being conducted. For example, a benefit from one perspective may be a cost from another perspective; a retrofit incentive is a benefit to the participant but is a cost to the utility. IWR-MAIN uses the following five accounting perspectives in evaluating a given conservation program.

1. *Participant:*   Benefits and costs are defined from the point of view of those utility customers who will participate in the program.

2. *Utility:*   The public or private water utility can be expected to face costs or receive program benefits that are different from the program participants.

3. *Community:*   This perspective takes the point of view of the local community, which includes both the utility and the program participant. However, this is not a simple summation of utility and participant cost and benefits. Nonparticipating ratepayers are also part of the community perspective.

4. *Society:*   The societal perspective considers benefits and costs that include, but also go beyond, the community perspective. Environmental benefits and costs and other externalities are included in this perspective.

5. *Ratepayer:*   The effect of a conservation program on water and/or wastewater rates is captured in the ratepayer's perspective. This perspective includes both program participants and nonparticipants.

IWR-MAIN uses a number of measurements to evaluate the economic merits of one or more conservation programs from each accounting perspective. The economic feasibility tests used are:

- *Net present value:*   The net present value (NPV) method provides a comparison of costs and benefits throughout the life of the conservation project. The NPV of a project is calculated as the difference between the present value of benefits and the present value of costs. A project with a positive (>0) NPV is economically viable.

- *Benefit-cost ratio:* For a given accounting perspective, this method determines the ratio of the present value of benefits to the present value of costs. Those conservation measures with a benefit-cost ratio greater than 1.0 are economically viable.

- *Discounted payback period:* The discounted payback method is used to estimate the consumer's perception of future benefits and costs. The discounted payback is the number of years it takes until the cumulative discounted benefits equal the cumulative discounted costs. Participants will prefer a shorter discounted payback period.

- *Levelized costs:* Levelized cost is a measure of total discounted cost of a conservation program per unit of water conserved; the cost is levelized over the life of the program.

- *Lifecycle revenue impact:* The lifecycle revenue impact (LRI) measures the direction and magnitude of a one-time change in water rates resulting from implementation of a particular conservation program. LRI will be positive for conservation programs that, over their lifetime, cause an increase in water rates.

### Summary

A complete description of the IWR-MAIN forecasting methods, socioeconomic input data requirements and options, conservation savings methods, and benefit-cost methods is presented in *IWR-MAIN, Version 6: Water Demand Analysis Software—User's Manual and System Description* (1996). Figure 4-2 summarizes the inputs and outputs of the IWR-MAIN Water Demand Analysis Software.

## Water Requirements Modeling with the Installation Water Resources Analysis and Planning System (IWRAPS©)

Water use at military installations can be viewed as similar to normal urban water uses in that the installation uses water for a variety of domestic, commercial, and industrial activities. However, forecasting future water use requires an understanding of the distinct differences between the military and civilian communities.

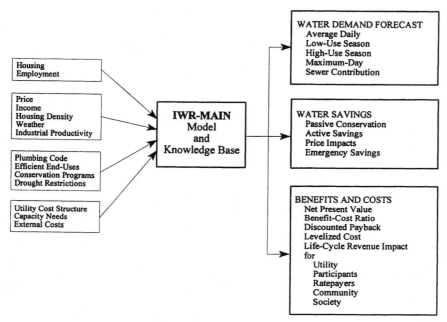

**Figure 4-2.** IWR-MAIN Inputs and Outputs

Military installations are like small cities in many respects. Installations have housing for married and unmarried personnel, dining facilities and restaurants, shopping centers, recreation centers, and medical facilities much like civilian cities. However, the main industry of military installations is often quite different from their civilian counterparts. The military operations at installations which can be very different from any civilian activities require constant training. Often the training is conducted at off-installation locations resulting in large temporary installation population decreases. In other cases, additional military personnel are assigned to an installation for training, resulting in sudden large increases in installation population.

Planning methodologies employed to estimate water use at military installations pose many interesting challenges. As a result of sudden installation population changes, water planners must be able to supply the increased water necessary to meet the needs of the installation. Often, increases in installation water requirements exceed or tax the existing water supplies. In these cases, planners must have ongoing or emergency water conservation measures to meet the increased requirements. In addition to the challenges of a rapidly changing water use level, the absence of water meters on individual buildings creates difficulties for installation water planners. Without meters, planners are often unable to identify where water is used or to assess the effects of any conservation measures imposed. To aid the installation water planner in forecasting water needs in times of sudden population increases (mobilization) and to estimate future water savings from the implementation of conservation measures, the Installation Water Resources Analysis and Planning System was developed.

IWRAPS© provides a procedure to determine sectorially disaggregated seasonal water requirement estimates at military installations in the continental United States. Individual forecasting models have been developed for the Army, Air Force, and Navy and Marine Corps. The models estimate the winter requirements based on the building square footage and activity level for each building sector. The square-footage methodology developed by Langowski et al. (1985) estimated annual water requirements by multiplying the water use coefficient in gallons per square foot per day (GPSFPD) by the square footage of buildings. The Langowski model utilized only three sectors and made no consideration for intensity of use. Thus, the method was unable to identify specifically where water was being used, seasonal differences in water use, or account for increased water use with increased installation population associated with mobilization.

In response to this need, the first version of IWRAPS© (Feather et al. 1993) was developed using a nationwide sampling of water use at fixed U.S. Army installations. When estimating water requirements, consideration was given to the activity level within 21 distinct water use categories. This method was also successfully applied to CONUS (continental United States) Air Force (Willett et al. 1995b) and Navy installations (Willett et al. 1995a). Its methodology assumes that installation water usage will be directly proportional to the activity level of the building sector. The Air Force model utilizes winter water requirements, weather inputs, square footage of water-using buildings, climatic zones, and the primary tenant activity (mission) of the installation to determine summer water requirements. The Navy model adds variables for the existence of golf courses.

All three IWRAPS© models incorporate future construction and demolition of buildings to predict varying future water requirements. To fit the needs of installation planners, IWRAPS© can be used to estimate restricted forecasts, given specific conservation measures, and/or mobilization plans. The potential effectiveness of conservation measures is evaluated using fractional reduction in water use and coverage resulting from each individual measure. Mobilization forecasts consider the impacts of personnel in field settings as well as additional water requirements within existing real property assets.

## Building Types and Sizes

In the IWRAPS© models, water use is directly related to building types and sizes that are surrogate variables for the mission and population of an installation (Langowski et al. 1985). A change in mission at an installation may be reflected by changing the building types and their respective sizes. Consequently, a change in water use at the installation will result. Based on each building type having a specific water use requirement, the model features 21 IWRAPS© sectors signifying 21 building types (Table 4-4). For example, the water requirements for warehouse buildings are different from those of family housing, based on different water-using activities in these two sectors. Family housing typically requires water for cooking, dish washing, laundry, and showering. Warehouse buildings might use water for cleaning the floors and minimal sanitary needs. The area of buildings in each sector is the principal driver of the model. It is assumed that building size is a primary determinant of sectoral water use.

In addition to the 21 basic IWRAPS© sectors, the system allows user-added special-purpose categories for facilities that may not be clearly identified into a single building sector or where water use is best explained in an alternative metric. A separate evaluation of special-purpose water use is performed in IWRAPS©. Examples in the special-purpose category are golf courses and vehicle wash facilities. These categories allow the user to specify the water use in terms of convenient coefficients, such as gallons per square foot per day, winter/summer gallons per day (GPD), or gallons per vehicle. Facilities such as golf courses can conveniently be represented in terms of winter/summer gallons per day if the necessary data are available. Other special-purpose uses may vary in type of unit-use measures, such as gallons per acre or each. Table 4-5 presents a list of the more commonly identified special-purpose water uses.

**Table 4-4.** Real Property Sectors in IWRAPS©

| | |
|---|---|
| Administration/operations | Guest housing |
| Reserves | Health/dental clinics |
| Barracks | Hospitals |
| Bowling center | Laundromat/dry cleaner |
| Banks/credit union | Maintenance |
| Bachelor officer quarters | Restaurant/cafeteria |
| Community buildings | Schools |
| Commissary | Service stations |
| Dining | Warehouse |
| Family housing | Exchange facilities |
| Gyms | |

**Table 4-5.** Common Special Purpose
Water Uses in IWRAPS©

| Swimming pools | Hot water facility |
|---|---|
| Vehicle wash facilities | Fire test centers |
| Aircraft wash facilities | Golf course irrigation |

## Construction and Demolition

IWRAPS© allows the user to include any available data regarding anticipated or planned changes in mission or sector building areas. These
changes can significantly affect water use in future years. Installations frequently undergo changes in building structures due to routine replacement
of older buildings or construction of new buildings to meet the requirements of mission changes. The robustness of the water-requirements forecast increases with more accurate future construction and demolition data.
It is likely that new structures will have more efficient water infrastructure
and will use less water per square foot than existing buildings. Therefore,
IWRAPS© allows for a change in the efficiency factor of new construction
in forecast years to reflect changes in building standards and efficiency
increases in water-using appliances.

## Seasonality

CONUS military installations generally exhibit two distinct seasons, with
winter running November through March and summer from April
through October. These seasons are easily differentiated by mean monthly
maximum temperatures in excess of 65 degrees Fahrenheit (Table 4-6).
High temperatures above this level generally signal an increase in outdoor
water use for such activities as lawn irrigation, outdoor cleaning, and
operation of swimming pools.

IWRAPS© uses two separate seasonal models to forecast annual water
use. The winter model estimates water requirements assuming primarily
indoor water use during this season. Thus, the winter model utilizes a multiplicative product of building area and a sector-specific water use coefficient to estimate overall sectoral water use. Special-purpose water use is
added to the sectoral water use total to provide a winter season forecast.

The summer model utilizes a statistical model relating winter water use,
mean monthly maximum summer temperature, mean monthly summer
rainfall, climatic zones, and the primary tenant activity of the base to total
summer requirements. In addition to these variables, the Navy model
includes total installation land area and the existence of a golf course. The
Air Force model employs the basic set of variables and an adjusted real

**Table 4-6.** Seasonal Variation in Temperature
and Precipitation

| | Mean temperature* | Mean maximum temperature* | Mean minimum temperature* | Mean precipitation[+] |
|---|---|---|---|---|
| Jan | 43.9 | 53.3 | 34.5 | 3.28 |
| Feb | 46.8 | 56.6 | 36.9 | 3.03 |
| Mar | 52.5 | 62.8 | 42.1 | 3.54 |
| Apr | 59.5 | 70.2 | 48.7 | 2.82 |
| May | 66.6 | 77.1 | 56.2 | 3.36 |
| Jun | 73.2 | 83.5 | 62.9 | 3.36 |
| Jul | 77.0 | 86.8 | 67.1 | 4.07 |
| Aug | 76.3 | 86.1 | 66.4 | 3.92 |
| Sep | 71.2 | 81.2 | 61.1 | 3.63 |
| Oct | 62.8 | 73.2 | 52.3 | 3.01 |
| Nov | 53.7 | 63.7 | 43.8 | 3.61 |
| Dec | 45.7 | 55.2 | 36.3 | 3.00 |
| Winter[+] | 48.5 | 58.3 | 38.7 | 3.29 |
| Summer[+] | 69.5 | 79.7 | 59.2 | 3.45 |

Period of record: 1945–1990 from naval installations in the continental United States.
* Temperature in degrees Fahrenheit.
[+] Precipitation in inches.
[+] Winter is November through March. Summer is April through October.

property area as a surrogate for improved installation acreage. The difference between total summer and winter seasonal water use is proportioned among the sectors based on winter water use ratios unless seasonal differences are known and input into the system.

**Climate**

The very long term weather patterns and the geography of a location combine to provide regional climate characteristics. The Köppen's climate classification system (Figure 4-3) provides an accepted method of identifying each of eight major climatic zones in the continental United States (Murphey 1977). Each zone is characterized by similar long-term weather patterns, vegetation, elevation, and other physical characteristics. These climatic characteristics provide valuable information about required water use within that region.

**Installation Mission**

Military installations are planned and constructed in order to accomplish specific assigned tasks or missions. Water use is highly dependent not only

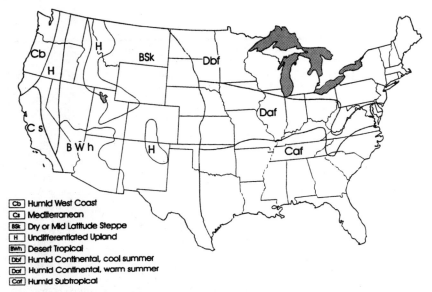

**Figure 4-3.** Köppen's Climatic Classification System

on weather, climate, and size of the installation, but also on how water is being used. Classification of military installations into similar mission groups allows the water planner to make some generalizations on how water is used at a particular installation. For example, the primary mission of an installation can range from administration, recruit training, maintenance training, or any of a variety of operational specialties. Installations primarily devoted to administrative activities would be expected to use substantially less water than an adjacent recruit training base of equivalent size. Similarly, medical and maintenance installations would have different water uses.

Each of the three IWRAPS© models includes categorical variables to represent the fixed effects of significant mission types. The Army and Air Force models categorize mission based on the major command (MAJCOM). Each MAJCOM has a specific mission type. Because the Navy is organized differently, the Navy and Marine Corps model required manual classification based on the primary tenant activity at an installation.

## Installation and Fixed Effects

The Air Force and Navy models include categorical variables for specific installations that capture installation-specific water use patterns and water use policies. For example, if an installation is not operated year-round, the categorical variable is used to adjust the installation's model intercept to reflect partial usage. In addition, the Navy model utilizes a categorical variable denoting the existence of a golf course. During analysis of Navy and Marine Corps water use, the existence of an installation golf course significantly impacted overall water use. Therefore, to individually capture and explain installation water use, a golf course variable was included in the model to represent the increased water use over and above the level required for normal irrigation.

## Conservation and Mobilization

IWRAPS© evaluates the effectiveness of conservation measures that reduce water use or water loss. The structure of conservation analysis follows the conservation model contained in the Institute for Water Resources Municipal and Industrial Needs (IWR-MAIN) Version 5.1 (Davis et al. 1991) used for water demand management in nonmilitary communities. IWRAPS© allows simultaneous evaluation of as many as six conservation measures. The effectiveness of a given measure depends on the fractional reduction in water use as a result of implementing the conservation measure, coverage of measure, and unrestricted water use in the IWRAPS© sector. The level of reported savings assumes that each evaluated conservation measure is independent of all other measures evaluated. If water savings in one measure affect potential savings in another measure, adjustments should be made to the fractional reduction in all affected measures. IWRAPS© also evaluates water use, given a planned mobilization scenario. It accounts for the change in water use due to the increased troop strength and additional construction, if any, necessary to accommodate the added population.

## Water-Requirement Algorithms

**Water Use Coefficients and Level of Activity Adjustments.**  Determination of water use coefficients plays a major role in IWRAPS©. Ideally, if an installation had 100 percent metering of *all* 21 separate building sectors, the IWRAPS© user could assign water use coefficients derived from the actual metered data to each IWRAPS© sector. Unfortunately, current metering is partial at best. IWRAPS© allows user selection of representative coefficients from a library of building sector coefficients developed from three service specific nationwide surveys of military installations. Mean water use coefficients generated from all service branches are presented in Table 4-7.

IWRAPS© incorporates adjustments for level of activity in each sector. For example, maintenance buildings at a time of higher level of activity

**Table 4-7.** Normalized Winter Water Use Coefficients
by Service Branch

| Sector name | Sector code | Mean water use coefficient (gpsfpd) | | |
|---|---|---|---|---|
| | | Air Force | Army | Navy |
| Administration/ operations | A | 0.06701 | 0.20915 | 0.07201 |
| Reserves/ National Guard | AR/NR* | 0.07645 | 0.05255 | 0.03146 |
| Barracks | B | NC[†] | 0.15611 | 0.32128 |
| Bowling center | BC | 0.07877 | 0.04341 | 0.21520 |
| Banks/credit unions | BK | 0.05404 | 0.08014 | 0.17496 |
| Bachelor officers quarters | BOQ | 0.20187 | 0.02500 | 0.11098 |
| Community buildings | C | 0.11589 | 0.07254 | 0.11493 |
| Commissary | COM | 0.08390 | 0.12082 | 0.07784 |
| Dining | D | 0.39486 | 0.23112 | 0.36740 |
| Family housing | F | 0.18372 | 0.22485 | 0.16721 |
| Gymnasium | G | 0.06526 | 0.14719 | 0.04792 |
| Guest housing | GH | 0.03278 | 1.82504 | 0.25472 |
| Health/dental clinics | HDC | 0.64599 | 0.12282 | 0.13505 |
| Hospital | HOS | 0.57075 | 0.76046 | 0.21313 |
| Laundromat | L | 0.24389 | 1.04664 | 1.16721 |
| Maintenance | M | 0.15946 | 0.26235 | 0.08137 |
| Restaurants | R | 0.33906 | 0.49680 | 0.22393 |
| Schools | S | 0.12245 | 0.03951 | 0.05564 |
| Service stations | SS | 0.58908 | 0.07842 | 0.11295 |
| Warehouse | W | 0.04630 | 0.02427 | 0.12108 |
| Exchange | X | 0.23553 | 0.18766 | 0.09630 |
| Totals | | 0.24131 | 0.28208 | 0.17266 |

* AR is the sector code for Air Force and Army reserves buildings, NR is the sector code for Navy reserve buildings.
[†] NC indicates no coefficient is currently available.

brought about by mission changes will require more water when compared with water quantities required for normal operations. It is assumed that the levels of activity at one installation will differ from activity levels of others. Also, at different periods of time, a specific installation may have different levels of activity. Fluctuations in activity level will be reflected in a change in water use at the installation. The forecasting system accounts for the level of activity by analyzing shortage or surplus of space at a given time. Shortage or surplus of building space is measured by the ratio of actual-versus-required square footage for each IWRAPS©

sector. Required square footage is determined by missions assigned or inferred for an installation. IWRAPS© uses real property reports as a source of information on required and actual square footage.

**Winter Model Algorithms.** The concepts discussed in this chapter, which affect the winter and summer water requirement estimates derived in IWRAPS©, are shown in the following mathematical form. The winter water requirement comprises five elements as follows:

$$Qw_j = \sum_{i=1}^{21} [AC_{ij} \cdot STAND_{ij} \cdot ADJCONS_{ij}] + SPEC_j + UNACC \qquad (5)$$

Though the base year and time steps are adjustable, this presentation follows the convention below:

$j = 0$ denotes the base year
$j = 1$ first forecast period
$j = 2$ next forecast period
and so forth.

The other variables describing winter water use ($Qw_j$ in GPD) are:

$i$ = water use sector
$j$ = time period of forecast
$AC_{ij}$ = actual square footage for sector $i$ in time period $j$
$STAND_{ij}$ = standardized* coefficient from sector $i$ in time period $j$, gallons per square foot per day
$ADJCONS_{ij}$ = tally of building construction adjusted for efficiency factor for sector $i$ in time period $j$
$SPEC_j$ = special-purpose water use in time period $j$
$UNACC$ = unaccounted water

The actual square footage ($AC_{ij}$) in the preceding equation is calculated as a successive increment of the value of $AC$ from the base year ($j = 0$):

$$AC_{ij} = AC_{i,j=0} + \sum_{k=1}^{j} [CONS_{ik} - DEMO_{ik}] \qquad (6)$$

where  $CONS_{ik}$ = building construction for sector $i$ during time period $k$, in thousand sq. ft.
$DEMO_{ik}$ = building demolition for sector $i$ during time period $k$, in thousand sq. ft.
$k$ = time period counter

---

* The normalization and standardization of winter water coefficients is described under coefficient development later in this chapter.

The standardization coefficient is developed using the present nomenclature,

where:     $RQ_{ij}$ = required square footage for sector $i$ in time period $j$
               $C_i$ = normalized water use coefficient for sector $i$

The required square footage term $(RQ_{ij})$ requires special handling to account for mission-related construction or demolition.

$$RQ_{ij} = RQ_{i,j=0} + \sum_{k=1}^{j} [(CONS_{ik} - DEMO_{ik}) \cdot M_{ik}] \qquad (7)$$

Here $M_{ik}$ is a mission flag for sector $i$ in time period $k$ and equals 1 if the construction or demolition is mission-related and 0 otherwise.

Inclusion of the new building efficiency factor requires an accounting of new-versus-original building square footage when calculating $ADJCONS$. This distinction process is shown as follows:

$$ADJCONS_{ij} = \frac{B_i + \sum_{k=1}^{j} \left[ CONS_{ik} \cdot \left( 1 - \frac{e_{ik}}{100} \right) \right]}{AC_{ij}} \qquad (8)$$

where:     $B_i$ = original building square footage remaining in sector $i$ in
               terms of the base year
           $e_{ik}$ = new building efficiency factor for sector $i$ in time period $k$

The original building square footage term, $B_i$, is expressed mathematically as:

$$B_i = AC_i - \sum_{k=1}^{j} DEMO_{ik} \qquad (9)$$

After substituting the two elements in Equation 5, mathematical simplification results in the following equation for winter season water use:

$$Qw_j = \sum_{i=1}^{21} \left[ \left[ B_{ij} + \sum_{k=1}^{j} \left[ CONS_{ik} \cdot \left( 1 - \frac{e_{ik}}{100} \right) \right] \right] \cdot \left[ \frac{RQ_{ij}}{AC_{ij}} \cdot C_i \right] \right] + SPEC_j \qquad (10)$$

where:  $AC$ is expressed by Equation 6
        $RQ$ is expressed by Equation 7
        $B$ is expressed by Equation 9

**Summer Model Algorithms.**  The summer water requirement for a given water use sector is calculated as a function of the winter water requirement, climate, mission, and installation land area. Each branch-specific model utilizes a statistical variation of the multivariate relationships:

Summer water use = winter water use + seasonal difference

Seasonal difference = $f$(weather, climate zone, mission)

Summer water use = $f$(winter water use, seasonal difference, installation specific binary variables)

The models generated for each service branch are similar in many ways (Table 4-8). For example, they all use mean maximum temperature and mean precipitation from the summer months as surrogate variables for summer irrigation, climate parameters to partially explain geographic water use differences, and an intercept term to adjust the model as an alignment mechanism. The installation's mission and installation specific calibration coefficients have also been used as explanatory variables to predict summer water use.

**Table 4-8.** IWRAPS© Summer Water Use Models

| Model Components | Air Force | Army | Navy |
|---|---|---|---|
| Dependent Variable: | QDiff* | QRatio[†] | QSummer |
| Independent Variables: | | | |
| Intercept term | Yes | Yes | Yes |
| Winter water use | No | No | Yes |
| Mean monthly max summer temperature | Yes | Yes | Yes |
| Mean monthly summer precipitation | Yes | Yes | Yes |
| Square footage of water use sectors | Adjusted[‡] | Total | No |
| Installation acreage | No | Improved[§] | Total |
| Köppen's climate classification zones | | | |
| Midlatitude steppe (BSK) | Yes | No | No |
| Subtropical desert (BWH) | No | No | Yes |
| Humid subtropical (CAF) | Yes | Yes | No |
| Humid west coast (CB) | Yes | No | Yes |
| Mediterranean (CS) | Yes | Yes | No |
| Humid continental cool summer (DAF) | No | Yes | Yes |
| Humid continental warm summer (DBF) | No | No | Yes |
| Undifferentiated upland (H) | Yes | No | No |
| Mission | Yes | No | Yes |
| Existence of golf course | No | No | Yes |
| Installation calibration coefficients | Yes | No | Yes |
| Adjusted model R[†] | 0.84 | 0.44 | 0.97 |

* Estimate of the difference between summer and winter water use.
[†] Estimated as the ratio of summer and winter water use.
[‡] Adjusted square footage includes all water use sectors except warehouse and maintenance.
[§] Improved acreage includes areas that receive annual funding for maintenance.

The model employed for estimating summer season water requirements at U.S. Navy and Marine Corps installations directly estimates the total summer season water requirement using the statistical model presented as follows:

$$QSUMMER_j = -32692905 + 1.466399Qwinter_j + 857389TEMP_j$$

$$- 2560849PRECIP_j + 279.390748AREA + 82005754BWH$$

$$- 14109107CB - 36730324DAF - 23706039DBF \qquad (11)$$

$$+ 39538081MAR - 158133862PRT - 12890201STA$$

$$+ 23699336SUB + 14443262TRN + 7108890GOLF$$

$$+ Installation\ Term$$

where:   $Qwinter_j$ = installation winter season water requirements in time
                period $j$, in gallons
     $TEMP_j$ = mean monthly maximum summer temperature in
                time period $j$
   $PRECIP_j$ = mean summer precipitation in time period $j$
      $AREA$ = total installation land area
      $BWH$ = climate zone code—subtropical desert
       $CB$ = climate zone code—humid west coast/maritime
      $DAF$ = climate zone code—humid continental, hot summer
      $DBF$ = climate zone code—humid continental, cool summer
      $MAR$ = primary tenant activity code—marine infantry
      $PRT$ = primary tenant activity code—port or shipyard
      $STA$ = primary tenant activity code—naval station
      $SUB$ = primary tenant activity code—submarine base
      $TRN$ = primary tenant activity code—training base
     $GOLF$ = golf course existence code
        $j$ = time period

The summer-to-winter seasonal water requirements difference, $QDIFF$, is then computed as the difference between Equation 11 and Equation 10.

$$QDIFF_j = QSUMMER_j - Qwinter_j \qquad (12)$$

Summer season forecasts for the U.S. Air Force employ a statistical model to estimate the seasonal difference ($QDIFF$) which is then added to the winter season estimate produced by Equation 10.

## Sectoral Allocation

Total summer season water use is allocated to each real property sector based on the winter water use proportion of the sector. The following relationship describes the assignments of the proportion of total $QDIFF_j$ for each individual water-using sector:

$$SECTPROP_{ij} = \frac{Qw_{ij}}{\sum_{i=1}^{21} Qw_{ij}} \qquad (13)$$

$$QDIFF_{ij} = SECTPROP_{ij} \cdot QDIFF_j \qquad (14)$$

$$Qs_{ij} = Qw_{ij} + QDIFF_{ij} \qquad (15)$$

where:   $SECTPROP_{ij}$ = winter water use proportion for sector $i$ in time period $j$
$QDIFF_{ij}$ = difference between summer and winter GPD for sector $i$ in time period $j$
$Qw_{ij}$ = winter water use for sector $i$ in time period $j$
$Qs_{ij}$ = summer water use for sector $i$ in time period $j$

The above algorithm produces a unique $QDIFF$ for each sector. The sum of all the individual difference values for each sector must always equal the $QDIFF$ value computed by the regression model. This relationship implies that there is a uniform proportionality in each sector between winter and summer water use. However, an installation having substantial meter data for a sector may find its $QDIFF_i$ in summer to be greater than or less than the predicted aggregate $QDIFF$. For example, metered data for a family housing sector may show that the $QDIFF_{FH}$ summer need is greater than what is reflected in the predicted $QDIFF$ value. Therefore, the computational procedure given below provides a mechanism for the planner to override the $QDIFF_{FH}$ for the family housing sector.

$$QDIFF_r = Qs_r - Qw_r \qquad (16)$$

Here, $r$ is the sector where $QDIFF$ is being overridden. Summer water use in sectors other than the overridden sectors (where $i \neq r$) must ultimately, however, be adjusted proportionally as well in order to maintain the aggregate $QDIFF$ value predicted by the model. The procedure for that adjustment is:

$$SECTPROP^*_{i \neq r} = \frac{Qw_{ij}}{\sum_{i \neq r}^{21} Qw_{ij}} = \frac{Qw_i}{Qw - Qw_r} \qquad (17)$$

$SECTPROP^*$ is the adjusted winter water use proportion for the sectors $i \neq r$.

$$QDIFF^*_{i \neq r} = (QDIFF - QDIFF_r) \cdot SECTPROP^*_{i \neq r} \qquad (18)$$

$QDIFF^{*}_{i \neq r}$ represents the adjusted difference between summer and winter water use for each sector ($i \neq r$) that is not overridden but requires recomputation to equal the original $QDIFF$ that represents the sum of all proportions.

## Conservation Algorithm

The effectiveness of a given conservation measure depends on the fractional reduction in water use as a result of implementing the conservation measure. The fractional reduction, or measure efficiency, is determined by the percent reduction expected from measure implementation and the portion of the sector affected by measure implementation. The IWRAPS© conservation effectiveness algorithm is as follows:

$$E_{mij} = R_{mi} \cdot C_{mij} \cdot Q_{ij} \tag{19}$$

where: $E_{mij}$ = effectiveness of conservation measure $m$ (e.g., plumbing retrofit) for sector $i$ (e.g., family housing) in time period $j$

$R_{mi}$ = fractional reduction in use of water for sector $i$ as a result of conservation measure $m$

$C_{mij}$ = coverage of measure $m$ used in sector $i$ at time $j$ expressed as a fraction

$Q_{ij}$ = unrestricted water use in sector $i$ at time $j$

The coverage factor $C_{mij}$ indicates the proportion of the sector to which the measure has been applied that will be affected by the measure being analyzed (measure $m$). The reduction factor $R_{mi}$ refers to the assigned fractional reduction that is estimated to result from the implementation of a conservation measure $m$ in sector $i$. The software for IWRAPS© allows only a single reduction value for each conservation measure analysis. Therefore, summer and winter reduction values must be evaluated as a weighted average for each season individually.

The overall reduction in water use through conservation impacts the unaccounted water sector as well. The percent of unaccounted water is considered as stable regardless of the total reduction due to conservation, unless the overall quantity of water use is reduced. If the total quantity of water use is reduced, as with the implementation of leak detection, the unaccounted water use sector will also be reduced.

Each of the water conservation measures evaluated provides the potential to decrease total installation water requirements. It is possible to implement one or more of these measures. When assessing the total savings associated with implementing a mix of these measures, total potential savings can be created by adding each measures's individual potential savings only when each measure is independent of the other. Whenever there is measure overlap in particular sectors, the effect of implementing overlapping measures differs from the additive total. Thus, it is reasonable to assume that the installation of the deluge bypass valve in the maintenance sector and the retrofitting of plumbing fixtures in family housing will provide an additive total potential water savings estimate.

However, in arid locations, the implementation of a landscape restriction measure and the installation of evapotranspiration (ET) controls on the landscape irrigation system shows considerable overlap. These measures are necessarily related. Whenever landscape restrictions return a particular area to its natural desertlike state, irrigation requirements are greatly reduced, thus reducing potential savings associated with the installation of the ET controls. In cases like this, the reduction factors must be recomputed to reflect the total potential savings taking into consideration the effects of the overlapping measures.

**Unaccounted Water Use.** According to the American Water Works Association (AWWA) unaccounted water is the difference between total water produced and the amount delivered to customers. In the context of a military installation, unaccounted water is the difference between total installation production and total installation consumption. Differences typically result from system losses attributable to leakage, uncalibrated metering, and activities like "firefighting, street washing, sewer flushing, and other unmetered public services" (Moyer 1985). Unaccounted water loss is a parameter common to all utilities. Studies have shown acceptable distribution losses of 10 to 20 percent (Keller 1986). Percent unaccounted water rate should be estimated based on system audit information and/or analysis of production records, distribution system description, and discussions with installation personnel. Unaccounted water is computed for all water entering the base distribution system.

### Forecast Procedures

The development of an installation water use forecast is accomplished utilizing a six-step process. These steps are detailed in the following sections.

**Data Acquisition and Review.** As with all water use studies, the first step in constructing a water use forecast is gathering data on past and present water use. In order to efficiently gather data, an appropriate base year should be selected. The selected base year should represent a normal water use year, that is, normal weather and average water use (no large leaks or extended emergency supply shutdowns). Water use levels for the base year should be available. Once an appropriate base year has been selected, sectoral water use records should be obtained for as many individual building or sector accounts as is possible. Nonsectoral, or special-purpose water uses such as vehicle and aircraft washes, swimming pools, and golf courses, should receive special attention. As the primary driver variable for the installation water use forecast, base year real property records (building area by type) in addition to all planned building construction and demolition should be obtained.

**Real Property Verification.** Real property records obtained in step one should be analyzed for accuracy. Attention should be devoted to verifying that installation real property records include the building area for contract or nonmilitary activities, such as restaurants, schools, banks, and credit unions. In many cases real property values for these activities must be added. Planned changes in installation real property via new construction, demolition, or reassignment of existing assets to other sectors should be detailed for each forecast year and included in the real property input modification screens by forecast year.

**Coefficient Development.** Using the individual building or sectoral water use billing records and information in the real property file, water use coefficients are developed. These coefficients represent an average water use per square foot of area per day during the winter season in each building sector. The coefficient is based on winter use because winter use is most often associated with indoor water use. While some outdoor use is included, it represents only minimal levels. A method used to generate raw water use coefficients is presented in Table 4-9.

**Table 4-9.** Raw Coefficient Development

|  | Water use (gallons) at the base exchange | | | |
|---|---|---|---|---|
| Month | 1993 | 1994 | 1995 | Mean |
| November | 723,800 | 630,900 | 658,900 | 671,200 |
| December | 851,000 | 656,500 | 370,400 | 625,967 |
| January | 651,800 | 629,500 | 291,400 | 524,233 |
| February | 700,500 | 585,600 | 810,700 | 698,933 |
| March | 764,900 | 625,600 | 412,850 | 601,117 |
| Winter Total | 3,692,000 | 3,128,100 | 2,544,250 | 3,121,450 |
| Building Area (sq. ft.) | 111,416 | 111,416 | 111,416 | 111,416 |
| Winter Days | 151 | 151 | 151 | 151 |
| Raw Coefficient | 0.219451 | 0.185933 | 0.151229 | 0.185538 |

Water use coefficient $(U)$ = winter water use/building square footage/days in winter
$U = 3,121,450/111,416/151 = 0.185538$ GPSFPD

Using a strict gallons per square foot per day approach would result in the same predicted water use for buildings that are fully occupied or totally empty. Therefore, the coefficients are adjusted to remove the effects of over- or underutilization based on the ratios of actual-to-allowable square footage values. The allowed square footage value for a facility (or sector) is a design parameter based on the number of military personnel required to perform the mission or task for which the facility was designed. Comparison of actual square footage to allowable square footage values allows the planner to determine whether a facility is overutilized, thus using more water than normal, or underutilized (or even empty).

The process of normalization estimates what the water-usage rate would be under normal activity levels. Normalization utilizes the ratio of actual to allowed square footage to "remove" the effects of over- or underutilization. All coefficients contained in the software have been normalized by the authors. New coefficients developed by installation water planners should be normalized prior to entry into IWRAPS©. An example of normalization is presented in Table 4-10.

Once normalized coefficients have been developed, they may be compared and utilized for other facilities in that sector or even at other installations. However, in order to import a coefficient derived at another facility or installation, the activity levels at the receiving installation must be reinserted into the coefficient. The process of reinserting the effects of higher or lower than normal activity levels is called *standardization.* Standardization utilizes the ratio of allowed-to-actual square footage to reflect the activity levels in water use rates. Table 4-11 presents an example of the standardization.

Water use rates for nonsectoral, or special-purpose uses, must also be computed on a gallons per day basis. However, because square footage may not be the appropriate metric for units, the software allows water use

**Table 4-10.** Coefficient Normalization

Water Use in Administration/Operations Sector

Actual size = 560,000 ft$^2$
Allowed size = 480,000 ft$^2$
Measured water use = 230,400 gpd

$U$ = 230,400 GPD/560,000 ft$^2$
= 0.4114 GPSFPD

Find $U'$, the normalized coefficient based upon:

$U'$ = U * Actual : Allowed
$U'$ = 0.4114GPSFPD * 560,000 ft$^2$/480,000 ft$^2$
= 0.4800 GPSFPD

**Table 4-11.** Coefficient Standardization

| Standardization of Coefficients |
| :---: |
| Commissary Sector |

Actual size = 8200 ft$^2$
Allowed size = 7500 ft$^2$
$U'$ (normalized) = 0.1472 GPSFPD

Find $U''$, the standardized coefficient based upon:

$U'' = U' *$ Allowed : Actual
$U'' = 0.1472$GPSFPD $* 7500$ ft$^2/8200$ ft$^2$
$U'' = 0.1346$ GPSFPD

The standardized water use estimate for the commissary is:

$0.1346$GPSFPD $* 8200$ ft$^2 = 1104$ GPD

rates per unit to be assessed in a variety of metrics (e.g., per acre, per square foot, per pool, each, etc.). Special-purpose water use rates must be computed for both winter and summer season. The use of different seasonal water use rate coefficients provides increased flexibility and accuracy for predicting special-purpose water use. For example, swimming pool and golf course irrigation water use are highly seasonal with a minimal water use rate during the winter, but a substantial contributor to total installation water use in the hotter summer months.

**Calibration.**   Once all required forecast data have been gathered, verified, and computed, it is entered into the software system using a series of data input screens. Before computation of all forecast years, it is advisable to calibrate the software to ensure reliable results. Calibration is performed by comparing predicted water use during the base year to the recorded use levels. By adjusting the level of unaccounted water use, the predicted base year water use should align with observed water use. The goal of the calibration step is to derive accurate annual water use predictions. Examination of seasonal forecasts may differ slightly depending on how the installation's seasonal water use compares to the expected seasonal split (seven-month summers/five-month winters). Once the model has been calibrated by iteratively selecting unaccounted water percentages to minimize the difference between predicted and observed base year water use, unaccounted water use should be left constant over the entire forecast period. If adjustments to percent unaccounted water fails to provide acceptable seasonal predictions (> ±5 percent), further study should be directed to verifying the completeness and accuracy of the input data. Table 4-12 provides the observed accuracy ratings of the three models. The

**Table 4-12.** Comparison of Base Year
Accuracies for IWRAPS© Applications

| Application | Winter | Summer | Annual |
|---|---|---|---|
| Army* | 99.9 | 100.4 | 100.1 |
| Air Force[†] | 102.0 | 102.3 | 102.3 |
| Navy[‡] | 104.3 | 97.6 | 100.0 |

* Six site applications.
[†] Five site applications.
[‡] Two site applications.

accuracy rating is the ratio of predicted to observed water use expressed as a percentage.

**Conservation Measure Selection and Analysis.** Military water resource planners need to consider water conservation in their overall master planning. This is linked to the realization of constraints to the development of additional water supplies, the Federal Energy Policy and Conservation Act of 1992, Executive Order 12902 (1994), and stronger requirements to meet environmental compliance regulations. Implementation of water conservation practices provide alternatives to water utility planners that can directly avoid these limiting factors and maintain compliance with regulations by providing improved efficiency in water use.

Water conservation planning with IWRAPS© uses a five-part planning process:

1. Selection of potential conservation measures

2. Targeting of sectors to implement conservation measures

3. Development of water use reduction values

4. Analysis of potential water savings

5. Economic evaluation of the cost effectiveness of selected conservation measures

The first four steps are integral parts of the IWRAPS© input and output. Economic evaluation of water conservation measures is achieved from simple economic procedures of savings and costs resultant of conservation efforts.

Conservation measures are ideally selected for their potential to reduce water use. Common water conservation measures that have been used at military installations include plumbing retrofit, leak detection and repair, public education programs, water recycling systems in aircraft and vehicle wash systems, xeriscaping, and landscape water use restrictions. Selected water conservation measures are part of an overall water conservation program.

Sectors to receive a conservation measure are targeted for their compatibility with a selected measure. For example, a shower retrofit program would be applicable in the family housing, barracks, bachelor officer quarters, guest housing, and gym sectors since showers are likely located in these buildings. In contrast, a public education program has the potential to reduce water use in all sectors including the special-purpose sectors. Likely sectoral coverage of selected conservation measures is presented in Table 4-13.

Water use reduction values are best based on empirically derived reduction values from similar measures implemented elsewhere. Reduction values are available in professional journals manuscripts, public agency study reports, and consultant reports. For example, Davis et al. (1991), Dziegielewski et al. (1986), the U.S. Army (1983), and numerous state agency reports are valuable sources for determining a water conservation reduction value. However, professional judgment on the part of the water utility planner who has knowledge of the installation's water use characteristics is ultimately the determining factor in selecting the best value to use for a selected measure and targeted sector. A list of ranges in reported reduction values are presented in Table 4-13.

Analysis of water savings occurs once conservation measure selection, sectoral targeting, and reduction values are determined. The potential water savings are computed in regard to unrestricted water use, or water that is required at an installation if no conservation efforts are undertaken. The potential water savings associated with measures should be compared to the cost of implementing conservation actions.

Various economic evaluations including a benefit-cost analysis and discounted-payback periods are useful to judge the cost effectiveness of conservation efforts. For example, a benefit-cost analysis simply compares the cost of implementing a measure and the savings associated with reduced water use. A 10-year discounted payback period has recently been used as a determining factor in funding DoD 1391 projects. Water and energy conservation projects that have a 10-year or less discounted-payback period are potential projects to undertake in accordance with Executive Order 12902 (1994).

**Mobilization Scenario Analysis.** Added water requirements under mobilization can be viewed as a unique consideration of military water utility planning. It is a very real possibility and should receive attention during utility design and emergency management. Military installations involved in mobilization contingencies are required to prepare "mobilization plans." IWRAPS© uses data from these plans to provide an estimate of water requirements under mobilization conditions.

**Table 4-13.** Reduction Values and Probable Sectoral Targets

| Water use sectors | Public education | Plumbing retrofit | Pressure reduction | Landscape restrictions | E.T. sprinkler Trigger |
|---|---|---|---|---|---|
| *IWRAPS© Sectors* | | | | | |
| Administration/operations | 0.03–0.05 | | 0.05–0.30 | | |
| Reserve | 0.03–0.05 | | 0.05–0.30 | | |
| Barracks | 0.03–0.05 | 0.20–0.40 | 0.05–0.30 | | |
| Bowling center | 0.03–0.05 | | 0.05–0.30 | | |
| Banks/credit unions | 0.03–0.05 | | 0.05–0.30 | | |
| Bachelor officer quarters | 0.03–0.05 | 0.20–0.40 | 0.05–0.30 | | |
| Community buildings | 0.03–0.05 | | 0.05–0.30 | | |
| Commissary | 0.03–0.05 | | 0.05–0.30 | | |
| Dining | 0.03–0.05 | | 0.05–0.30 | | |
| Family housing | 0.03–0.05 | 0.20–0.40 | 0.05–0.30 | 0.50–0.58 | 0.05–0.50 |
| Gyms | 0.03–0.05 | 0.20–0.40 | 0.05–0.30 | | |
| Guest housing | 0.03–0.05 | 0.20–0.40 | 0.05–0.30 | | |
| Health/dental clinic | 0.03–0.05 | | 0.05–0.30 | | |
| Hospital | 0.03–0.05 | | 0.05–0.30 | | |
| Laundry/dry cleaning | 0.03–0.05 | | 0.05–0.30 | | |
| Maintenance | 0.03–0.05 | | 0.05–0.30 | | |
| Restaurant/cafeteria | 0.03–0.05 | | 0.05–0.30 | | |
| Schools | 0.03–0.05 | | 0.05–0.30 | | |
| Service stations | 0.03–0.05 | | 0.05–0.30 | | |
| Warehouse | 0.03–0.05 | | 0.05–0.30 | | |
| Exchange facilities | 0.03–0.05 | | 0.05–0.30 | | |
| *Special-Purpose Sectors* | | | | | |
| Aircraft wash | 0.03–0.05 | | 0.05–0.30 | | |
| Vehicle wash | 0.03–0.05 | | 0.05–0.30 | | |
| Golf course | 0.03–0.05 | | 0.05–0.30 | | 0.05–0.50 |
| Swimming pool | 0.03–0.05 | | 0.05–0.30 | | |

During full mobilization, many factors cause the buildup time and the total mobilization period to vary. The primary determination of mobilization requirements is the degree of mobilization taking place. It is recommended that conditions of full mobilization be evaluated. A buildup period of about 180 days takes place during which the magnitude of troop buildup and training is increased and planned mobilization construction takes place. Beyond the buildup period, an equilibrium point will be approached, and all designed assets will be constructed. A steady influx of new trainees will continue, but soldiers will also be leaving as they are deployed to the war zone. Water requirements must be forecast based on the following assumptions:

- Mobilization can occur at any time, and therefore, the additional water requirements needed to sustain the total mobilization force for a minimum of one year regardless of start time is the ultimate forecast objective.

- Mobilization troops and support personnel will be housed and serviced by three sectors: (1) existing structures, which will require use intensities to be adjusted based upon the proportional increase in the actual-to-allowed ratio; (2) new construction, which will drive use intensities to full utilization for the building and special-purpose sectors receiving the added square footage; and (3) field bivouac sites, where soldiers are expected to utilize about 150 gallons of water per day (GPD).

- Mobilization water needs will increase during the short term (initial 180 days) and stabilize during the long term (beyond 180 days). Plans should be developed for long-term needs, which facilitate and support the mobilization mission.

- Although mobilization can occur at any time, the amount of additional water must be coupled with unrestricted or restricted (conservation) peacetime water use projections for various time horizons to determine total installation water requirements.

The water requirements for those housed in new or existing buildings are estimated using building square footage. All sectors requiring construction under mobilization are assumed to operate at maximum-design mobilization level; thus the activity level is 1.0. Water requirements for the remaining sectors are assumed to increase proportionally to the mobilization impact or coverage on applicable sectors and to the added installation population attributed to mobilization.

Water requirements for the portion of population housed in the field are calculated using 150 gallons per capita per day (Smith 1984; U.S. Army 1984). This portion of the mobilization population should be netted out of the calculation of nonmobilization construction sectors. In some cases, such as Fort Bliss, it is planned that all mobilization personnel will eventually be accommodated in existing or new buildings. All water requirements are therefore based on an assessment of the resultant building areas.

The first phase or short-term planning consideration prorates the long-term mobilization water requirements to reflect water needs during the buildup. This exercise is directly dependent on the rate of mobilization construction completion. Potential temporary water shortages and other important demand management issues can be pinpointed through this type of analysis.

To summarize, the following nine steps should be followed in estimating water requirements under mobilization:

1. Determine additional mobilization troop population as a fraction of the premobilization troop population. This fraction is referred to as the *mobilization factor.*

2. Identify building-related sectors which require additional construction.

3. Determine the fractional impact of mobilization on all water-using sectors where new construction is not required. This fraction is referred to as the *mobilization coverage.*

4. Estimate the percent of additional mobilization troop population to be housed in bivouac setting in the short and long term.

5. Calculate water requirements for sectors identified in step 2 based upon an activity ratio of 1.0.

6. Calculate water requirements for sectors not identified in step 2 by multiplying baseline unrestricted water use by the mobilization factor and coverage values and adding the product to the premobilization unrestricted water use.

7. Calculate the field component water requirement based upon 150 GPCD.

8. Add steps 5–7 to obtain total mobilization requirement.

9. Calculate conservation effectiveness and the resulting reduction in water use based on the results of step 8.

The following equation summarizes the mathematical operation of the mobilization module.

$$QM_j = \left[ QM_F(FLD_t) + \sum_{i=1}^{21+N} QM_{i,j} \right] \tag{20}$$

where: $QM_j$ = total mobilization water use, GPD, for either summer or
winter

$FLD_t$ = fraction of mobilization troops in field setting at
mobilization stage $t$

$QM_{i,j}$ = water use under mobilization for sector $i$ in time period $j$,
GPD, for either a winter day or a summer day

$N$ = the number of special purpose sectors

$QM_F$ = water use by troops in field setting, GPD
$= TRP \cdot Q_F$

where: $TRP$ = number of arriving mobilization troops

$Q_F$ = water use per field troop, 150 gallons/troop/day

New construction required by mobilization can be identified by sector and is represented by the term $MCONS_i$. If $MCONS_i > 0$ for any building or special purpose sector, then

$$\sum_{i=1}^{21} [AC_{i,j} + MCONS_i] \cdot \left[ \frac{B_{\emptyset i} + MCONS_i}{AC_{i,j} + MCONS_i} + \frac{\sum_{k=89}^{j} CONS_{i,k}}{AC_{i,j} + MCONS_i} \cdot \left(1 - \frac{ei,k}{100}\right) \right]$$

$$\cdot \left[ \frac{AL_{i,j} + MCONS_i}{AC_{i,j} + MCONS_i} \cdot C_{Ni} \right] + SPEC \quad (21)$$

$$\sum_{i=1}^{21} [AC_{i,j} + MCONS_i] \cdot \left[ \frac{B_{\emptyset i} + MCONS_i}{AC_{i,j} + MCONS_i} + \frac{\sum_{k=89}^{j} CONS_{i,k}}{AC_{i,j} + MCONS_i} \cdot \left(1 - \frac{ei,k}{100}\right) \right]$$

$$\cdot \left[ \frac{AL_{i,j} + MCONS_i}{AC_{i,j} + MCONS_i} \cdot C_{Ni} \right] \cdot QRATIO_j + [SPEC \cdot QRATIO_j] \quad (22)$$

If $MCONS_i = 0$ for any building or special purpose sector, then

$$QWM_{i,j} = [(QW_{i,j} \cdot MFAC_{ij} \cdot MCOV_{ij}) + QW_{ij}] \quad (23)$$

$$QSM_{i,j} = [(QS_{i,j} \cdot MFAC_{ij} \cdot MCOV_{ij}) + QS_{ij}] \quad (24)$$

where: $MCONS_i$ = mobilization construction for water use sector $i$ in
square feet

$QW_{ij}$ = winter unrestricted water use for sector $i$ in time
period $j$, GPD

$QS_{ij}$ = summer unrestricted water use for sector $i$ in time
period $j$, GPD

$QWM_{i,j}$ = winter mobilization water use for sector $i$ in time
period $j$, GPD

$QSM_{ij}$ = summer mobilization water use for sector $i$ in time period $j$, GPD

$MFAC_{ij}$ = mobilization factor, which is the fractional increase in troop strength caused by added mobilization troops for sector $i$ in time period J

$MCOV_{ij}$ = mobilization coverage, which is the fractional value of expected impact on unrestricted water use for sector $i$ in time period J

The remaining variables were defined in this section on IWRAPS©.

The main inputs required in the IWRAPS© software are the change in troop strength by the added mobilization population, new construction, buildup rate, and the mobilization factor and coverage values for each sector. These inputs may vary, but this optional analysis can provide useful planning information under many specified mobilization scenarios.

For installations that have prepared a program to implement specific conservation measures across time, a mobilization event could occur simultaneously during the execution of the conservation program. If the program is achieving expected reductions in water use, it can tend to attenuate the impact of mobilization which requires more water. It can be assumed that arriving mobilization troops will be subject to the active and ongoing conservation program as well.

Although water utility managers must plan for the worst-case scenario, that is, no active water conservation program, it appears that the potential reductions in water loss and waste could enhance the mobilization effort, especially at installations where water could be a limiting factor. Savings attributed to conservation effectiveness could reduce, if not completely satisfy, mobilization water needs.

Computation of conservation effectiveness ($E$) under mobilization is identical to the procedures previously described with the exception that the unrestricted sectoral water use quantity ($Q$) includes the added water needs to sustain mobilization. In other words, conservation effectiveness is determined after the computation of total mobilization water requirements. Field water use by troops in a bivouac setting is assumed to be unaffected by ongoing water conservation. The adjusted algorithm is represented by the following:

$$E_{mij} = R_{mi} \cdot C_{mij} \cdot QM_{ij} \qquad (25)$$

where: $QM_{ij}$ = mobilization water use for sector $i$ in time period $j$, GPD, for either a summer or a winter day

It should be noted that the special purpose sector, referred to as unaccounted water, is subject to fluctuations caused by mobilization and/or conservation, or both. However, the percent of total water use attributed to this sector is maintained throughout each forecasted period, regardless of the scenario. The percent can change only if specific conservation measures are applied to obtain such a decrease, or if installation utility managers realize that greater water losses are occurring across time.

## References

Braun, L., P. L. Schaffer, and A. McMillin. 1994. *NAS Fallon.* San Diego, CA: Marcoa Publishing, Inc.

Baumann, D. D., J. J. Boland, and J. H. Sims. 1980. *Evaluation of Water Conservation for Municipal and Industrial Water Supply—Procedures Manual.* IWR Contract Report 80-1, Fort Belvoir, VA: Institute for Water Resources, U.S. Army Corps of Engineers.

Baumann, D. D., J. J. Boland, J. H. Sims, B. Kranzer, and P. H. Carver. 1979. *The Role of Conservation in Water Supply Planning.* Report 78-2, Fort Belvoir, VA: Institute for Water Resources, U.S. Army Corps of Engineers.

Davis, W. Y., D. M. Rodrigo, E. M. Opitz, B. Dziegielewski, D. D. Baumann, and J. J. Boland. 1991. *IWR-MAIN Water Demand Analysis Software—Version 5.1—User's Manual and System Description.* IWR Report 88-R-6. Fort Belvoir, VA: Institute for Water Resources, Water Resources Support Center, U.S. Army Corps of Engineers.

Department of the Navy. 1982. *Facility Planning Criteria for Navy and Marine Corps Shore Installations.* NAVFAC P-80. Washington, D.C.: Facilities Engineering Command.

Derhgawen, U. K., H. P. Garbharran, J. F. Langowski, Jr., C. A. Strus, and T. D. Feather. 1992. *Part II—Current and Projected Water Use and Conservation Effectiveness at Vandenberg Air Force Base.* Carbondale, IL: Planning and Management Consultants, Ltd.

Derhgawen, U. K., H. P. Garbharran, J. F. Langowski, Jr. 1993. *Water Requirements at Fort Huachuca Military Reservation.* Carbondale, IL: Planning and Management Consultants, Ltd.

Dziegielewski, B. and C. Strus, 1993. *Managing Urban Water Demands.* American Water Works Association Proceedings of 1993 Annual Conference. Water Resources Volume. pp. 243–251.

Dziegielewski, B., E. M. Opitz, J. C. Kiefer, and D. D. Baumann. 1993. *Evaluating Urban Water Conservation Programs: A Procedures Manual.* Denver, CO: American Water Works Association.

Dziegielewski, B., E. M. Opitz, W. Y. Davis, and D. Baumann. 1986. *Water Conservation Evaluation for the Phoenix Water Service Area. Volume 1: Technical Report.* Carbondale, IL: Planning and Management Consultants, Ltd.

Energy Policy and Conservation Act. P.L. 92-163, 89 Stat 871, 42 U.S.C. 6201 et seq.

Executive Order 12902. 1994. *Energy Efficiency and Water Conservation at Federal Facilities. Federal Register,* vol. 59, no. 047.

Feather, T. D., C. A. Strus, R. E. Robinson, J. F. Langowski, Jr. 1993. *Volume II—Installation Water Resources Analysis and Planning System (IWRAPS©).* Carbondale, IL: Planning and Management Consultants, Ltd.

Keller, C. W. 1976. "Analysis of Unaccounted-for Water." *Journal of the American Water Works Association,* vol. 68, no. 3, pp. 159–62.

Langowski, J. F., Jr., J. E. Lang, J. T. Bandy, and E. D. Smith. 1985. *A Survey of Water Demand Forecasting Procedures on Fixed Army Installations.* CERL Technical Report N-855/07. Champaign, IL: Construction Engineering Research Laboratory, U.S. Army Corps of Engineers.

Moyer, E. E. 1985. *Economics of Leak Detection: A Case Study Approach.* Denver, CO: American Water Works Association.

Murphey, R. 1977. *Patterns on the Earth: An Introduction to Geography.* Chicago, IL: Rand McNally College Publishing Company.

Perdue, B. L. and H. Garbharran. 1993. "Special-Purpose Water Requirements Sector Use at Air Force Installations: Aircraft Wash Facilities." In Willett, J. S., N. A. Hanna-Somers, R. J. Hauer, and J. F. Langowski, Jr. 1995a. *Installation Water Resources Analysis and Planning Systems—Navy (IWRAPS©–NAV) Development.* Carbondale, IL: Planning and Management Consultants, Ltd.

Perdue, B. L. and H. Garbharran. 1993. "Special-Purpose Water Requirements Sector Use at Air Force Installations: Vehicle Wash Facilities." In Willett, J. S., N. A. Hanna-Somers, R. J. Hauer, and J. F. Langowski, Jr. 1995a. *Installation Water Resources Analysis and Planning Systems—Navy (IWRAPS©–NAV) Development.* Carbondale, IL: Planning and Management Consultants, Ltd.

Planning and Management Consultants, Ltd. 1996. *IWR-MAIN Water Demand Analysis Software, Version 6.1: Users Manual and System Description.* Carbondale, IL.

Planning and Management Consultants, Ltd. 1996. *IWR-MAIN Water Demand Analysis Software: Technical Overview.* Carbondale, IL.

Smith, E. D., J. T. Bandy, W. P. Gardiner, and F. Huff. 1984. "Closed-Loop Concepts for the Army: Water Conservation, Recycle, and Reuse." CERL Technical Report N-85-85/0. Champaign, IL: Construction and Engineering Research Laboratory, U.S. Army Corps of Engineers.

Tetra Tech, Inc. 1994. *Environmental Assessment for the Management of the Greenbelt Area at Naval Air Station (NAS) Fallon, Nevada.* Prepared for Western Division, Naval Facilities Engineering Command. San Francisco, CA: Tetra Tech, Inc.

U.S. Army. 1986. *Water Conservation Methods for U.S. Army Installations: Volume 1, Residential Usage Management.* Technical Report N-146. Champaign, IL: Construction Engineering Research Laboratory.

U.S. Department of Army. 1984. Water Supply, General Considerations: Mobilization Construction. EM-1110-3-160. Washington, D.C.

Willett, J. S., N. A. Hanna-Somers, R. J. Hauer, and J. F. Langowski, Jr. 1995a. *Installation Water Resources Analysis and Planning Systems—Navy (IWRAPS©–NAV) Survey and Development.* Carbondale, IL: Planning and Management Consultants, Ltd.

Willett, J. S., N. A. Hanna-Somers, and J. F. Langowski, Jr. 1995b. *Installation Water Resources Analysis and Planning Systems—Air Force (IWRAPS©–AF) Development.* Second Edition. Carbondale, IL: Planning and Management Consultants, Ltd.

Willett, J. S., N. A. Hanna-Somers, and J. F. Langowski, Jr. 1995c. *Historical Water Requirements at Mountain Home AFB, Idaho.* Carbondale, IL: Planning and Management Consultants, Ltd.

# 5

# Price and Rate Structures

## W. Michael Hanemann
### University of California at Berkeley

This chapter deals with the principles for pricing water delivered to municipal users by public water supply systems. In principle, there are many different rate structures that could generate the same total revenue for a public water utility. The question, therefore, is what type of rate structure to adopt. In this chapter we consider some of the criteria that have been proposed for designing rate structures, we review the economic theory of how rates should be set, and we examine some of the empirical experience with different types of rates in the United States.*

## Components of a Water Rate Structure

Earlier this century, most urban water agencies were financed through fixed monthly charges. While these charges did not vary directly with consumption, they did sometimes vary according to customer characteristics, which, in the absence of metering, utilities used to identify probable consumption levels. In 1907, for example, Phoenix charged $1.50 per month for domestic water supply to dwelling houses; 25¢ per month each for

---

* This is adapted from Mitchell and Hanemann (1994), which can be consulted for a fuller discussion of the issues raised here.

water for horses and cattle; $1.00 for barber shops without fixtures, $1.50 for barber shops with one basin and one chair, and 50¢ for each additional chair; and rates of $1.00–$4.00 for stores and $2.50 and up for saloons. Since those days, water rates have changed, becoming simpler in some ways, more complex in others. The common feature of most water rates today is that they are based on metered use.

There are several basic components common to most water rate structures which we briefly discuss here. We first make a distinction between rates based on *flat charges* that are entirely independent of the quantity of water used (e.g., the monthly service charge of $1.50 per dwelling unit) versus charges that *vary* in some manner with the quantity used (e.g., 50¢ per 100 cubic feet). Regarding the latter, one can further distinguish charges that vary *directly* with the quantity of water used versus those that vary only *indirectly* or approximately with water use (e.g., a monthly service charge based on meter size or based on the number and type of fixtures). During the course of the century, many urban water agencies have switched from a system of fixed charges that were partly flat charges and partly based on factors that varied roughly, but imperfectly, with usage to a system that may include a flat charge, but also will include a variable charge based on metered usage.

A second important distinction is that between variable charges that are uniform rate versus block rate. A *uniform-rate* variable charge is one where the amount paid per unit of consumption is the same over all units consumed; a *block rate* is one where the unit charge varies, either decreasing with the amount consumed (decreasing-block rate) or increasing with the amount consumed (increasing-block rate). To the extent that a utility's fixed monthly charge allows some usage for free, this implies a form of increasing-block rate, even if the remaining volume charge does not increase with additional usage. In other words, the first block is priced at zero while the second block is priced at some positive amount. Until about 1980, decreasing-block rates used to be common for large, nonresidential accounts, if not for residential accounts, but are now giving way to uniform and increasing-block-rate structures.

Rates may also be divided between those that charge different prices in different time periods versus those that charge the same price in all time periods. In the water industry it is increasingly common to observe rates that vary by season; volume charges are higher during the peak season and lower during the off-peak season. These are referred to as *seasonally differentiated rates*, or more simply as *seasonal rates*.

In addition, water agencies may levy other charges not yet mentioned. An example is a one-time charge levied when new customers are connected to the water system, known variously as a *connection charge, facilities charge*, or *capacity charge*. This is intended to recover capital expenditures for new facilities required to meet the projected demands of new customers. Furthermore, many water agencies have some special rates for particular classes of user. For example, there may be *life-line rates* which offer low-income customers some initial quantum of usage at a reduced price. It is also common practice for water agencies to offer service to some users outside their service area at special, higher rates; or to offer water for large irrigation users on an interruptible basis at specially reduced rates.

A rate structure is built using a combination of the basic elements just discussed. The number of possible design permutations is almost endless. One finds rates that are uniform; uniform but adjusted seasonally; increasing-block; decreasing-block; increasing- or decreasing-block *and* seasonally adjusted; so on and so on. How does one choose? What type of rate design is best? In the next section we review some of the criteria for designing a rate structure.

# Criteria for Designing Water Rates

Generating revenue to meet a utility's revenue requirements is the primary role of a water rate structure. But intended or not, any system of water rates also performs two other functions: it allocates costs among users and it provides incentives to users. To evaluate the effectiveness of a given rate structure, we must consider how well it performs these three basic functions.

1. *Generate revenue*—rates generate revenue that permits a utility to cover its costs.

2. *Allocate costs*—rates serve to allocate costs among different types of use and user.

3. *Provide incentives*—rates provide price signals to customers which may serve as incentives for them to use water efficiently, encouraging them to modify their behavior in particular directions.

An effective rate structure, therefore, should do a good job of generating revenue, allocating costs, and providing incentives for efficiency. What is meant in each of these cases by "doing a good job"? What are the criteria for success? We next identify and comment on the basic criteria that are generally accepted in the literature.

## Criteria for Revenue Generation

With regard to revenue generation, a rate structure should be judged along at least three dimensions. Each represents a facet of the utility's ability to operate on a self-sustaining basis and meet its current and future financial obligations.

1. *Revenue sufficiency*—the rate structure should provide revenues sufficient to allow the utility to operate on a self-sustaining basis. The rate structure should generate revenue sufficient to cover operating costs such as salaries, chemical supplies, gas and electricity, taxes, and capital costs for system expansion, upgrades, or equipment replacement.

2. *Net revenue stability*—the rate structure should be conducive to a stable and predictable stream of net revenue over time and as circumstances change (e.g., in a drought). Stable net revenues allow for more accurate budgeting, better planning, and can lower long-term financing costs. Conversely, unstable net revenues increase the risk of insufficient cash flow and can raise long-term financing costs. To the extent that cost does not automatically vary in line with revenue, instability in revenue can generate instability in net revenue.

3. *Administration*—administration and billing costs should be balanced against the potential benefits of a more complex rate structure. The rate structure should be designed to accommodate rate modifications and updates.

## Criteria for Cost Allocation

With regard to cost allocation, the two principal considerations are equity and economic efficiency. Equity involves at least two considerations:

4. *Fair allocation of costs*—the rate structure should apportion costs of service among the different uses and users in a manner that is fair and is not arbitrary.

5. *Avoid cross subsidies*—the rate structure should apportion costs of service in a manner that avoids the subsidy of one group of users at the expense of another.

In the context of cost allocation, economic efficiency requires the following:

6. *Fully allocate private and social costs*—the cost allocation created by a rate structure should fully reflect the private and social costs of providing service for the different users and uses that the utility serves.

## Criteria for Providing Incentives

With regard to incentives, the criteria reflect the goals of economic efficiency, system load management, and the promotion of conservation:

7. *Static efficiency*—the rate structure should encourage the efficient use of water in terms of quantity used and timing of use. To this end, at the margin, the water rate should reflect the full private and social marginal cost of supply.

8. *Dynamic efficiency*—the rate structure should encourage an efficient pattern of growth in water use and an efficient pattern of system development over time. In this regard, the marginal rate should reflect the long-run rather than the short-run marginal cost of water supply.

9. *Conservation*—the rate structure should provide proper incentives for conservation, including investment by water users in cost-effective water-saving appliances, fixtures, and landscaping.

10. *Rate structure transparency*—the rate structure should be easy for water users to understand so that it provides a clear price signal to them. In order to be effective at influencing their investment decisions, it should offer a predictable price signal with minimum risk of unexpected adjustments.

## Applying the Criteria in Practice

Applying these criteria to determine the best rate structure in any particular situation is not necessarily an easy task, for at least three reasons. The first is that some of the criteria may directly conflict, forcing water agency managers to make trade-offs among them. For example, there can be a conflict between revenue stability and economic efficiency. Since a large part of a water utility's costs are fixed in the short run, the safest way to ensure revenue stability is to raise revenues entirely through a fixed monthly service charge. This totally insulates revenues from fluctuations in the quantity of water delivered, but is counterproductive from the perspective of efficiency because it provides no incentive to use water sparingly. Conversely, some degree of revenue instability is inherent in any rate structure with volume charges.

The second reason, which builds on the first, is that more than one party is involved in the rate-setting process and, typically, different parties place different weights on the alternative criteria because of their own particular interests and point of view. Beecher et al. (1990) describe the rate-making process as "a continual balancing act among the divergent and often competing perspectives of utilities, consumers, and society." They characterize these perspectives through a series of questions which we reproduce here, matching them up with the criteria presented in the preceding section.

## The Utility's Perspective

- Does the rate structure fully compensate the utility so that revenue requirements are met? (Criteria 1, 2)
- Does the rate structure allow the utility to earn a fair return on its investment? (Criteria 1, 2)
- Is the rate structure strategically sound for load management, competition, and long-term planning? (Criteria 1, 2, 7, 8)

## The Customer's Perspective

- Are both the rate-making process and the rate structure equitable? (Criteria 4, 5)
- Are utility rates perceived to be affordable? (Criterion 10)
- Are both the rate-making process and the rate structure understandable? (Criterion 10)

## Society's Perspective

- Does the rate structure promote economic efficiency? (Criteria 6, 7, 8)
- Does the rate structure promote the appropriate valuation and conservation of resources? (Criteria 6, 7, 8, 9)
- Does the rate-making process take into account priority uses of water? (Criterion 9)
- Are both the rate-making process and the rate structure just and reasonable? (Criteria 1, 4, 5)

Because of these differences in perspective, and because many parties have a stake in the rate-setting process, one should recognize that rate design in practice is an inherently political process.

The third reason why applying the criteria can be difficult is the complicated nature of water utility costs. In many cases, the costs of service do not vary with the quantity of service in a simple or direct manner. The water industry is highly capital intensive, even compared to most other utility industries. Many of its expenses are fixed costs that hardly vary with the amount of water delivered. Moreover, the capital investment is generally long-lived, and the capital stock employed in any given water system tends to include assets of several different vintages, acquired at different times and at vastly different costs. Because capacity cannot readily be altered at short notice while demand may be highly variable—both by day and by season—a water utility always needs to build some amount of excess capacity into its system just to be sure of meeting demand with an acceptable level of reliability. Who is responsible for these costs and what price will a utility's customers be willing to pay for reliability? To complicate matters further, parts of a utility's system often serve several different functions at the same time, so that there is a problem of allocating joint costs among separate beneficiaries.

The result is that it can be both difficult and frustrating to answer what people might think is a simple question: "How much does it cost to supply me with the water that serves my needs?" On the one hand, it may require an enormous amount of data and extensive technical analysis. On the other hand, even when the analysis has been completed, there still may be no single, correct way to develop a rate structure that satisfies everybody. The very nature of water utility costs tends to create conflicts between the criteria of raising adequate revenues, allocating costs fairly, and providing incentives for efficiency and conservation. Indeed, it is disagreement over how to resolve those cost-driven conflicts that accounts for most of the major controversies over utility rate making. This is particularly true of the dispute between proponents of the traditional approach to rate making, based on what is known as the principle of *embedded cost,* and proponents of the newer approaches, associated with *marginal cost* pricing, that were first introduced into the electricity industry in the 1970s, then spread to other regulated industries such as natural gas and telecommunications, and are now being introduced into the water industry. The traditional approach pools many different costs into a single average. It looks backward to the costs that the utility has already incurred, and emphasizes the estimation and allocation of historical (i.e., embedded) average cost. In terms of the criteria presented earlier, it places the main emphasis on criteria 1 through 4. The newer approaches place primary emphasis on sending users the right price signal about the scarcity value of the commodity. They are generally forward looking: as a basis for setting prices, they look to marginal cost rather than average cost, and replacement cost rather than historical cost. In terms of the criteria, they place the most weight on criteria 6 through 9.

Given that differing notions of cost underlie the alterative approaches to rate making in the water utility industry, it is useful to review the economic concept of costs and the economics of pricing, to which we now turn.

## Water Supply Costs and Complexities

We start off by imagining a fairy tale water utility that obtains its water supply from a magic spigot deep in a forest. The spigot is guarded by elves who also are responsible for installing and maintaining the distribution network that serves the utility's customers. For all these services, the elves charge the water company $0.05 per hundred cubic foot (CCF) of water delivered to the utility's customers. The staff of the utility consists of one retired elf, who deals with all the customers, sends them bills for water service, collects their payments, and handles all the dealings with the elves who operate the magic spigot. Since he already has a pension, he performs these services for free.

In these circumstances, rate making is simple—everybody would surely agree on $0.05/CCF as the appropriate retail charge for water. What makes the story a fairy tale is not the cheap price for water but rather the simple cost structure. The water agency incurs costs only as and when water is delivered. The costs vary directly with the quantity of water delivered. All units of water delivered cost the agency exactly the same amount of money.

The real world is different in every respect. The water industry is highly capital intensive; most of its costs are capital costs, incurred when capital assets are installed, rather than operating costs, incurred as water is delivered. Once installed, the capital is not particularly malleable: the capacity cannot be quickly expanded if you suddenly need more; nor can it be disposed of profitably if you need less. Economists make a basic distinction between *fixed* versus *variable* costs—that is, costs that don't vary with the quantity of service provided versus those that do. Thus, most water utility costs are fixed costs. However, there are several complications to the distinction between fixed and variable costs. One arises from differences in the time frame of the analysis. We just indicated that, because capital is relatively unmalleable once installed, its cost is to be regarded as fixed. However, when viewed from the longer time perspective, all capital is variable—existing capital will wear out and, with sufficient time, new capital can always be added. This corresponds to the economists' distinction between *long-run* and *short-run*—in the short run, the capital stock is fixed and its cost is a fixed cost; in the long run capital is variable and its cost constitutes a variable cost. How, then, should water be priced? Should its price reflect a short-run or long-run perspective? If prices reflect short-run costs, this may encourage patterns of water use that are poorly suited to future circumstances when more expensive sources will be required. On the other hand, if prices reflect long-run costs, the utility may be in the position of charging for facilities before they exist.

Economists also make a basic distinction between *average cost* and *marginal cost*. Average cost is simply total cost divided by the quantity sold (i.e., it is the cost per unit); marginal cost is the change in cost per unit change in quantity sold (i.e., it is the increment in cost per unit increase in quantity, or the decrement in cost per unit decrease in quantity). When all units of a commodity cost the same, as in the fairy tale example, there is no difference between average and marginal cost—the two costs are equal.

Otherwise, however, there is a difference between average and marginal cost, sometimes a very large difference. Moreover, the distinction between average and marginal cost applies in both the short run and the long run. Thus, there are short-run average and marginal costs when capital is in place and its cost sunk, and long-run average and marginal costs, when capital is adjustable and all costs are variable.

Differences between average and marginal cost can arise for many reasons. An important example of divergent short-run average and marginal costs is the case of a utility with a fixed capital stock and a constant unit-operating cost. In the short run the marginal cost is simply the unit-operating cost—the cost of producing an additional unit. The average cost, however, consists of the unit-operating cost plus the cost of the fixed equipment averaged over the amount of production. The more capacity is used, the more product over which to spread the fixed cost and, therefore, the lower the average cost per unit. Hence, there is a declining short-run-cost curve lying above a short-run marginal cost curve, as shown graphically in Figure 5-1. Historically, the notion illustrated here—that with increased production one can spread fixed costs and therefore one has declining-average costs—underlies many of the arguments for declining-block prices that were used prior to the 1970s.

A more difficult case for a utility is when cost is falling in the short run but rising in the long run, as illustrated in Figure 5-2. This phenomenon has to do with the fact that new supplies may be considerably more expensive to develop than past supplies. While a utility may experience economies of scale for a single project—similar to the situation shown in Figure 5-1—it may not experience economies of scale over a range of pro-

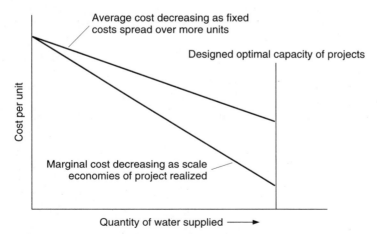

**Figure 5-1.** Short-Run Cost of Supply for Single Project

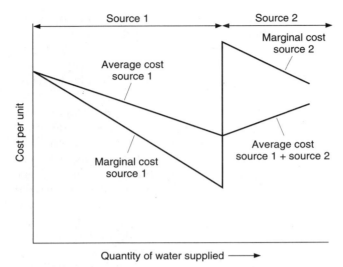

**Figure 5-2.** Change in Cost Structure Following Addition of New Supply

jects, meaning that cost is increasing as new capacity is added to the system. Consider a utility developing a supply of water. It secures the most convenient and least costly source available and builds a reservoir. With the investment made and the reservoir built, the cost of the water depends on the level at which the reservoir is utilized. As use of capacity increases the average cost of supplying a unit of water falls because the fixed cost is being spread over and more units. The short-run marginal cost of supplying a unit of water may also be falling because of scale economies associated with the project. When the capacity of the first reservoir is fully utilized, however, the utility will need to develop a new source of supply to meet new demand. It will likely have to develop a source not quite as conveniently located or of poorer quality than the first.

Suppose the cost of the second source is much higher than the first. How should the utility factor these two conflicting cost perspectives into its rate structure? Should rates reflect the lower unit cost of the first source, the higher unit cost of the second source, or some sort of average cost of both sources?

If we consider a utility with multiple sources of supply of varying vintages and costs, we get a situation like that shown in Figure 5-3. If the cost of water from these sources is arrayed from least to most expensive, as illustrated in the figure, we get an increasing long-run marginal cost of supply. In this case, the average cost would lie below the marginal cost; while rising, it is not rising as fast as marginal cost because it blends the

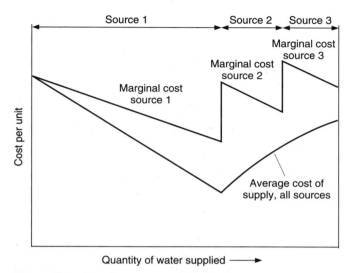

**Figure 5-3.** Cost Structure with Multiple Sources of Supply

high and the low costs together. This illustrates the complexity underlying the seemingly simple question, "How much does it cost to supply me with the water that serves my needs?"

This is the fundamental dilemma of rate making when average and marginal costs diverge—it is, in effect, an allocation exercise with many possible solutions. One approach is to average all costs over all users. Another is to charge all users the marginal cost for at least some of their usage. The first approach may be considered simpler, but it is not necessarily less arbitrary. Economic theory supports some version of marginal cost pricing on the principle that *all* users draw on the system at the margin and should be signaled the scarcity value of the resource. Whenever average and marginal costs differ, if all units are priced at the average cost then total revenue automatically covers total cost. This is not true when all units are priced at the marginal cost—total revenue falls short of marginal cost when marginal cost lies below average cost (Figure 5-1), but exceeds total cost when marginal cost lies above average cost (Figures 5-2 and 5-3). However, one cannot conclude that utilities always break even with average-cost pricing or that they can never break even with marginal cost pricing. The reason is that we have focused so far only on variable charges while most rate structures in practice, as indicated at the beginning of this chapter, have several components, including fixed charges. The revenues from other charges can supplement the revenues from marginal cost pricing in the situation depicted in Figure 5-1. Conversely, when the situation is like Figure 5-2 or 5-3, where marginal cost pricing would more than break even, one can hold revenues down in other ways—reducing some component of the fixed charge, for example, or charging less than marginal cost for part of a customer's usage (as with increasing-block rates).

In the next section, we discuss the economic case for marginal cost pricing in terms of presenting an appropriate signal to users of the commodity. We should point out, however, that determining this cost is often very complicated. The complexity of a water system will require that the utility identify and measure not just one but several marginal costs. Some of a water utility's costs vary with the number or type of customers rather than the amount of water delivered. Other costs, such as for system capacity, depend on the time as well as the amount of water use.

Unfortunately, the complications do not end here. Many of the facilities for a utility provide several services simultaneously, and hence the cost for those facilities may be joint or common to those services. If a facility can be expanded to increase the level of output for one service without affecting the provision of the other services connected to it, this does not raise any special issue. But, if the facility automatically benefits the provision of multiple services, then the costs of that facility or production process are said to be joint. For example, a distribution main jointly provides base capacity for off-peak service and extra capacity for peak service. To increase capacity by one unit for peak service necessarily increases capacity by one unit for off-peak service. In other words, producing one service implicitly results in the production of the other service, and their two costs are inextricably linked. As a result, the only economically definable cost for the distribution main—average, marginal, or otherwise—is for the composite peak/off-peak service it provides. This does not mean, however, that it is entirely impossible to derive prices for the separate services that are both efficient and recover their joint costs of production. But, it considerably complicates the pricing of these separate services and it requires prices not based on production costs alone.

To summarize, as a matter of principle one wants water rates to reflect costs. However, there usually is not one cost but many costs. Which costs, then, should rates reflect? While all the costs are real enough, it is how we classify them that matters for rate making. Any classification is a matter of judgment and, to some degree, inherently arbitrary. In the tale about the elves there was only one cost. Therefore, questions of classification did not arise and there could be total agreement on the correct rate structure. That is why it is a fairy tale.

## Marginal Cost Pricing

In this section we describe some of the approaches to pricing utility services that first emerged in the United States in the electricity industry during the 1970s and 1980s. What these approaches have in common is that they set aside the sole, narrow focus of meeting the utility's revenue requirements in favor of a broader set of goals that also include economic efficiency and the promotion of conservation.

Pricing to encourage more efficient use of water rests on the assumption that prices can change consumers' behavior, even for a basic commodity like water. Whether this is so is an empirical matter that certainly can vary with circumstances. A major point we wish to emphasize, however, is that *how* prices are used matters every bit as much as *whether* they are used. Prices can be effective or ineffective as tools for influencing behavior depending on how they are deployed.

The relative importance of price and other factors as influences on water use is frequently debated among water industry professionals. Our own view is that this is an unnecessary dichotomy—price matters, but so do other activities such as public education, advertising, and information-dissemination campaigns. Indeed, the two complement one another. Raising prices without providing guidance to customers on methods whereby they might reduce their water use blunts the effectiveness of the price signal. Conversely, exhortation and cajolery unaccompanied by financial incentives may have limited impact. All too often we have heard water managers complain that customers are resisting appeals for conservation because it is not in their financial interest. Low prices at the margin are an impediment to conservation—if there is little money to be saved, this undercuts the case for changing one's behavior.

Thus, we view pricing not as a substitute for a utility's existing or planned conservation programs but as something intended to work in tandem with them and enhance their impact. The hallmark of newer approaches to rate making, as we view it, is that they attempt to target or tailor prices in such a way as to create effective incentives to use water more efficiently.

## The Economic Argument for Marginal Cost Pricing

The economic argument for marginal cost-based rates comes from a branch of economics known as "welfare economics" which deals with prescriptions for efficiency in the use of scarce resources. It offers rules for the efficient total production of commodities, the efficient allocation of this total among individual producers and consumers, and the efficient expansion of production capacity over time through new investment.

The central prescription of welfare economics is that commodities should be produced and allocated to the point where their marginal benefit equals their marginal cost. Whenever this is violated, there cannot be full efficiency in the economy with respect to either total level of production or the allocation of this total among individuals. Marginal benefit is defined as the extra benefit from producing and consuming one more unit of the commodity; as explained in the previous chapter, marginal cost is the extra cost from producing and consuming one more unit of the commodity.* They can be equally well defined as the benefit lost and cost saved by producing one less unit; seen this way, marginal cost is synonymous with avoided cost—the cost that would be saved (avoided) by reducing output by a small amount. Seen the first way, it is synonymous with incremental cost—the added cost of a small amount of additional output.

A related concept that is very important to an understanding of resource allocation is *opportunity cost*. With a given productive capacity, a decision to produce more of any one commodity, say water for urban use, implies a decision to produce less of all other commodities. Thus, the cost to society of producing anything consists, really, in the other things that must be sacrificed to produce it. In the final analysis, all cost is opportunity cost— that is, the value of the alternatives foregone.

What is the economist's rationale for marginal cost pricing? Given a fixed productive capacity at any point in time, a decision to produce more of any one good implies a decision to produce less of all other goods as a whole. Therefore, the basic challenge for an economy is to use these resources to maximum advantage. For that to happen requires that the benefit gained from consuming one more unit of a good equal the cost to produce it. In other words, that the marginal benefit equal the marginal cost. How does one ensure this outcome? By ensuring that prices of commodities are set equal to their marginal cost. Prices provide the signals that guide people's behavior. When consumers make purchase decisions, they balance their benefit from a commodity against its price. If they are to make choices that reap the greatest possible benefit from society's limited resources, the prices that they pay must accurately reflect the opportunity costs associated with the commodities they are considering. If their judgments are correctly informed in this way, they will, by their independent decisions, guide scarce resources into those lines of production that maximize the net benefit to society *as a whole.*[†]

---

* Economists distinguish between *private* and *social* marginal benefit and cost. Social marginal benefit (cost) is the extra benefit (cost) accruing to society as a whole. Private marginal benefit (cost) is the extra benefit (cost) accruing to the individuals directly involved. To the extent that there are what economists call *externalities*, there can be a divergence between private and social benefits and costs. The modern economic theory of pollution, for example, grows out of the divergence between private costs (which exclude the effects of pollution emissions on other people) and social costs (which include the effects of these emissions).

[†] It is important to emphasize the "as a whole" in this statement. Economic concepts of efficiency remain quiet with regards to distributional concerns. It is quite possible to have an efficient allocation that is grossly inequitable. The question here becomes should equity concerns be addressed independent of or simultaneously with efficiency concerns. Economists generally favor that the two problems be addressed independently, though this prescription by no means enjoys unanimous consent.

Could not prices equal, say, average rather than marginal costs (assuming the two are different) and do this just as well? If price had no influence at all on demand, it would not necessarily matter. But, in fact, the demand for all commodities is in some degree, at some point, responsive to price. As a practical matter, we know this to be the case for water, as we noted in Chapter 2. Then, if consumers are to decide wisely from society's point of view whether to take somewhat more or somewhat less of any particular item, the price they pay must reflect the cost of supplying somewhat more or somewhat less—in short, the marginal-opportunity cost. Suppose, instead, that buyers were charged more than the marginal cost for a particular commodity; they would then buy less than the socially optimum quantity—welfare could be improved if they consumed more of the good in question and less of all other goods as a whole. Some consumers who would have consumed more of the good in question and less of other goods will refrain from doing so because the price exaggerates the good's opportunity cost. Conversely, if price is set below marginal cost, they will buy more of the commodity (and less of all other commodities taken together) than is socially optimal. Producers are diverting more resources to the production of this commodity than customers would have willingly authorized, had the price fully reflected the marginal-opportunity cost.

**The Choice of Time Frame for Marginal Cost Pricing.**    Although the basic economic argument for marginal cost pricing may seem compelling, we cannot end the discussion there and simply state that price should equal marginal cost. To do so is to ignore several important qualifications and practical limitations to this general rule. When it comes to moving from the theory of marginal cost pricing to its implementation, as with anything else, complications arise, and one is forced to deal with myriad details. In this and the next sections, we focus on two major issues: how one decides on the marginal cost concept to be used as a basis for setting prices, and how one ensures that the overall rate structure brings in sufficient revenue to meet the utility's requirements.

We earlier presented a brief tutorial on the economic concepts of cost and pointed out the distinction between short-run and long-run marginal costs, as well as the related distinction between fixed and variable costs. Fixed costs, that is, the costs associated with fixed inputs, are costs that do not vary with the quantity of service provided. Variable costs, that is, the costs associated with variable inputs, are the costs that *do* vary with the quantity of service provided. By definition, the marginal cost associated with the use of fixed inputs is zero—the quantity of output changes, but there is *no* change in the quantity of these inputs, and so there is no change in this component of total cost. Only variable inputs generate nonzero marginal costs. In the short run, the capital stock is fixed, and the marginal cost arises mainly from O&M costs. In the long run, the capital stock can be replaced and expanded, and the marginal cost includes not only O&M costs but also capital costs.

By definition, then, short-run marginal cost is always less than long-run marginal cost. But, this gap is especially huge in the water industry because of its unusually high capital intensity.* Thus, it matters greatly in the water industry whether prices are set on the basis of short-run marginal cost, which is only a small fraction of total expenditure, or long-run marginal cost.

What is to be done? Different opinions have been expressed in the economics literature, and there clearly is no easy answer. All agree that rice should never be set below short-run marginal cost. The question is whether it should be set any higher. The argument for short-run marginal cost pricing is best illustrated through the example of airline pricing. As Kahn (1988) has put it, no airplane should ever take off with empty seats as long as there exist some potential travelers who would be willing to pay the (almost negligible) short-run marginal cost associated with adding them to the flight roster. It is economically inefficient in this case to charge anything more than short-run marginal cost.[†]

This is not, however, a hard-and-fast rule. There are other considerations that may lead to a different conclusion. One consideration is price volatility. In some circumstances, short-run marginal cost could vary greatly over a short period of time as a result of fluctuations in either demand or supply (for example, in the case of a producer with access to many supply sources with very different costs). The variation in price that would result from short-run marginal cost pricing might be considered undesirable or counterproductive. Consequently, suppliers might prefer to abandon short-run marginal cost pricing in order to smooth out prices over time.

Another consideration is the impact setting price to short-run marginal cost may have on long-run investment. Whereas economic efficiency with

---

* For the water industry nationally, the asset requirement per dollar of revenue (i.e., the ratio of capital assets to annual revenue) is estimated at about $10–12; this is three to four times the capital intensity of the telephone and electric utility industries, and about five to six times that of the railroad industry.

[†] The airlines actually used this principle when they introduced cheap standby fares in 1996. But airline fares also reflect other economic considerations that are not necessarily socially efficient, such as price discrimination. The airlines reckon that the business demand for travel is far more inelastic than private individuals' demand for vacation travel. In order to discriminate between the two markets and maximize their profit from each, the airlines typically require staying over a Saturday night for their cheapest fares, since they figure this shuts out most business travelers. Then again, for standby fares to be efficient requires the airlines to be able to discriminate between different types of demand. Many airlines were compelled to discontinue standby fares because they found that standby customers were making false reservations under assumed names to ensure adequate seating.

respect to production from a given capital stock calls for setting prices equal to short-run marginal cost, economic efficiency with respect to investment and the determination of long-run capacity calls for setting prices equal to long-run marginal cost. If people are to reap the greatest possible benefit from society's limited resources and the capital stock is likely to be replaced or expanded in the foreseeable future, the prices of commodities must reflect the opportunity costs associated with not only the variable inputs but also the capital assets that are needed to produce them.

Suppose a producer will replace or expand his capital in the foreseeable future. Or, suppose that his customers will replace or expand their capital in the foreseeable future. In either case, charging short-run instead of long-run marginal cost provides an incorrect economic signal. If the wrong signal is given, this can lead both consumer and producer astray. If the price does not accurately reflect long-run marginal cost, the consumer may make investment decisions that are socially undesirable because they entail a long-run commitment to the use of a commodity for purposes that he would not consider justified if he had to pay the full, long-run cost of producing it. Likewise, the producer may make investment decisions that are socially undesirable because they entail a long-run commitment to the supply of a commodity whose users would not find it worthwhile if they had to pay the full, long-run marginal cost. This, in fact, is what happened on an unprecedented scale to the Washington Public Power Supply System—whose acronym WPPSS is appropriately pronounced *whoops*—resulting in the largest municipal bond default in U.S. history, which to this day continues to have serious financial repercussions for the Pacific Northwest.

Such questions about the correctness of investment decisions when prices are set below long-run marginal cost also have been raised throughout the history of the California water industry by critics of both urban and agricultural water policies. Gardner (1982) and others have long condemned the (wholesale) prices charged by the U.S. Bureau of Reclamation for this very reason. From the beginning, the Central Valley Project (CVP) has charged prices that were below long-run marginal cost—indeed, for most of the last two decades it charged prices that were below short-run marginal cost. As a result, the critics argue, it continued to build new reservoirs and aqueducts that irrigators would not have considered worthwhile had they been required to bear the cost. These investment projects were politically viable only because their costs were effectively invisible—they were averaged together with the costs of existing facilities and were spread over all beneficiaries of CVP water. But these phenomena—pricing below long-run marginal cost and making investment decisions for new reservoirs that would not be justified if those who used the additional supply actually had to bear the cost—are not unknown among urban water agencies, either.

As these considerations demonstrate, whether price should reflect short-run or long-run marginal cost will vary with the circumstances. As a general principle, however, we are inclined to follow Kahn's recommendation:

> The practically achievable benchmark for efficient pricing is more likely to be a type of average long-run incremental cost computed for a large, expected incremental block of sales, instead of short-run marginal cost estimate single additional sale. This long-run incremental cost (which we shall loosely refer to as long-run marginal cost) would be based on (1) the incremental variable costs of those added and (2) estimated additional capital costs per unit for the additional capacity that will have constructed if sales at that price are expected to continue over time or to grow. Both of these components would be estimated as average some period of years into the future. (Kahn 1988)*

Note that Kahn's use of "average cost" refers to the average incremental cost from adding a large block of new supply, *not* the average cost of supply from both old and new sources, which is the embedded-cost approach discussed in the previous chapter. When a utility prices its water service as recommended, customers are sent a signal that reflects the opportunity cost associated with their consumption level.

**Meeting Revenue Requirements with Marginal Cost Pricing.** We previously noted that, when prices are set equal to average cost, total revenue by definition covers total cost exactly. If prices are set in any other way, such as marginal cost pricing, this is not necessarily true. But we also noted two qualifications. One is where average cost per unit stays constant as output changes; the average and marginal cost coincide, so that marginal-cost pricing is the same as average-cost pricing. The more important qualification is that there are other charges, and even other ways of configuring marginal-cost pricing, through which utilities can ensure that their total revenue is adequate but not excessive to cover their total cost. Elaborating on these methods for meeting revenue targets is the focus of this section.

It has long been recognized that industries with decreasing average costs would not cover their total cost if they set prices equal to marginal cost. Two of the three main solutions have been known for over 50 years. One, involving what became known as the *two-part tariff,* was suggested by Lewis (1941). This is where a utility combines a commodity charge

---

* This recommendation serves as a general principle rather than a hard-and-fast rule. It is easy to imagine situations where pricing at short-run marginal cost would be most appropriate. For example, if a wholesale agency had extra water available in its reservoirs that it either had to sell or release to make storage available for next year's runoff, the appropriate price would be the short-run cost of delivering the water to retail water agencies.

based on marginal cost with a fixed charge, for example, a service charge or connection charge. Together, these make up the two parts of the tariff. The idea is that the fixed charge raises the additional revenue needed to cover total cost, but does not interfere with economic efficiency generated by having a commodity charge based on marginal cost. Since the customer pays the same fixed charge regardless of what quantity he consumes, only the commodity charge should influence his quantity decision, and this still provides the economically correct price signal. Hence, it is argued, the two-part tariff can satisfy the goals of both economic efficiency and revenue adequacy.

Subsequently, the efficiency of the two-part tariff has been challenged on two grounds. One is the argument that, even though consumers ought to disregard the fixed charge when deciding how much of the commodity to use, perhaps they do not. Perhaps they are impressed by the fact that, if they increase their consumption, they can spread this fixed charge over more of the commodity and so reduce the overall unit cost. If so, this would create an economically inefficient incentive to maximize consumption. When Mayor Bradley's Blue Ribbon Committee was meeting to consider LADWP water rates in 1992, it was persuaded by this argument and recommended eliminating fixed charges from the water rates. It is not clear, however, that the argument is valid.*

The second efficiency argument is given more weight by most economists. It arises from the fact that, while the fixed charge does not vary with the quantity consumed, it does vary with whether one consumes anything at all. If a consumer were willing to opt out of using the commodity entirely—for example, disconnecting from the utility's distribution network in order to avoid the fixed charge. A consumer in this position should be influenced by the fixed charge in making an economic decision. In this case, it can be shown that the fixed charge may be economically inefficient—there may be situations where the social optimum requires that the consumer not exit the system, but his private incentive in the face of the fixed charge does lead him to exit the system. In those situations, social efficiency would prescribe a smaller fixed charge than that needed to fill the gap between total costs and the revenues raised through marginal cost-based commodity charges. If so, there is no solution that both generates adequate revenues and is perfectly efficient from an economic point of view; there are only what economists call *second-best efficient* solutions.

---

* Another reason for having some fixed charge is to cover costs that vary by customer rather than by quantity of commodity sold, such as the costs of administering customers' accounts. Economic efficiency suggests these be recovered not by a commodity charge but by a fixed service charge set equal to the marginal cost per customer served.

Before continuing, we should emphasize that all the foregoing discussion is predicated on there being decreasing long-run average costs. In the urban water industry today, the reality is more likely to be increasing long-run average costs, in which case long-run marginal-cost pricing generates total revenues in excess of total costs. There is still an efficiency issue—the utility would need to dispose of the excess revenues in a way that doesn't bias customers' decisions away from efficiency—but it may be much easier to resolve. For example, the utility could use some type of rebate that was independent of customer use. Furthermore, the efficiency problems that arise from having a fixed charge (or rebate) would vanish if one could be sure that the charge did not influence customers' exit/entry decisions. For water, more than for any other utility service, this is likely to be the case. The fixed payments that most urban water agencies need to charge are unlikely to drive consumers off the distribution system and, therefore, are unlikely to cause a two-part tariff to be economically inefficient.*

The second main solution to the revenue problem of marginal-cost pricing in the presence of nonconstant average costs is known as *Ramsey pricing*. Ramsey (1927) modified the conventional economic efficiency analysis by adding an explicit constraint that commodity prices not only maximize social welfare but also break even.† He derived a complex formula for how one should adjust prices away from marginal cost in inverse proportion to the elasticity of demand. The intuition underlying Ramsey's formula is that, in order to preserve as much efficiency as possible, one wants to depart as little as possible from the pattern of consumption that would occur with unfettered marginal-cost pricing, while still charging prices that secure the utility sufficient but not excessive revenue. One accomplishes this goal by imposing the least price adjustments on the customers whose quantity demanded is least sensitive to price, and the smallest adjustments on the customers whose demand is most sensitive to price. The result is a form of cross-subsidization that yields a more efficient economy than if one had simply adjusted the price for all customers in the same way. There are two obvious problems with this approach, however. One is that the inverse elasticity pricing formula can be extremely complex and usually will require information on demand

---

* This is not the case with some other industries, such as cable TV for residential customers or long-distance telephone for large commercial and industrial customers. Those industries are more likely than water to have decreasing average costs, and they face a real prospect that their fixed charges could affect customer's entry/exit decisions. AT&T, in particular, has actively sponsored research since the late 1970s to develop a more efficient alternative to the two-part tariff known as *nonuniform* or *nonlinear pricing* [see Brown and Sibley (1986, Chaps. 4, 5), Mitchell and Vogelsang (1991, Chap. 5), and Wilson (1993)].

† Because of the imposition of this constraint, his analysis was an exercise—indeed, the *first* exercise—in second-best efficiency.

that simply is not available to most utilities. In order for Ramsey pricing to work, one must have detailed information not just for the demand of the commodity in question, but for all potential complementary and substitute goods as well. The other is that this particular solution to the revenue requirement problem relies on cross-subsidization, which tends to conflict with the principle of equity, and may be a violation of the regulatory codes governing the pricing of water.*

There are other, ad hoc solutions to the revenue problem created by divergent marginal and average costs. For example, Brown and Sibley (1986) note that in Europe it is sometimes accepted that public utilities should price on the basis of marginal cost with the government making up revenue shortfalls out of tax revenues. In the United States, the tradition has generally been that utilities cover their own costs without taxpayer assistance in the case of decreasing long-run marginal cost, or that they do not earn excess revenues in the case of increasing long-run marginal cost. This has generally led to average-cost pricing rules in the former case and some form of proportionally scaled-down marginal cost-based pricing in the latter. (The latter approach avoids the undesirable cross-subsidization implicit in Ramsey pricing by applying a uniform upward or downward scaling to marginal cost-based prices so that revenue equals total cost.) Either of these approaches distorts the price signal to the consumer away from the true long-run marginal cost.

Another solution is to have increasing- or decreasing-block rates. These can be regarded as multipart extensions of the two-part tariff. We mention them here only to point out their implications for revenue adequacy. Suppose there are decreasing long-run average costs. If there is a form of declining-block rates with the block where most consumption is located being priced at long-run marginal cost, while the earlier blocks (called the *inframarginal blocks*) are priced at some higher amount, this can provide many of the efficiency benefits of marginal cost pricing while still breaking even. Conversely, if there are increasing long-run marginal costs (which we consider more likely to be the case for the urban water industry), one wants a form of increasing-block rates with the inframarginal blocks priced below long-run marginal cost in order that total revenues match total costs.

**Marginal Cost Pricing Summary.**　Rate structures based on marginal-cost pricing precepts are intended to provide price signals that result in a more efficient allocation and use of a scarce supply of water. Efficient con-

---

* In other words, by forcing those least able to adjust their pattern of consumption to bear the brunt of the effort to meet a revenue constraint, Ramsey pricing sacrifices equity at the altar of efficiency. For more on this, see Baumol.

sumption requires that the benefit derived from consuming one more unit of a good equal the cost of supplying it. If the benefit is less than the cost, it is inefficient to produce additional units; the resources required to do so could be more productively employed elsewhere in the economy. By setting price equal to marginal cost, consumers are able to compare the benefit of additional consumption with its associated cost and make efficient choices. In this way, production may be guided toward more efficient levels.

Marginal-cost pricing differs from average-cost pricing in important ways. Whereas marginal-cost pricing reflects the cost of producing an additional unit, average-cost pricing reflects the unit cost of producing all units. If costs are decreasing as output increases, average cost will be above marginal cost. If costs are increasing as output increases, average cost will be below marginal cost. In either case, prices based on average cost will not result in efficient consumption choices because consumers will not be matching marginal benefit with marginal cost.

Marginal-cost prices are more difficult to calculate than average-cost prices. To calculate average-cost prices, a utility needs some understanding of its total costs and production. Dividing one by the other yields an average price. To calculate marginal cost, on the other hand, a utility needs fairly detailed information on costs for a variety of plants and equipment, some of which may not yet be built. This data may be expensive and difficult to obtain and of uncertain reliability.*

Marginal-cost pricing can result in over- or undercollection of revenue. Generating revenue insufficient to cover cost obviously is not sustainable in the long run. Collection of too much revenue may also be problem if a utility is constrained to earn zero profit. Various strategies to satisfy the zero-profit constraint while retaining the efficiency properties of marginal-cost pricing have been proposed. Two-part tariffs and multiple block rates, for example, can be designed so that price at the margin reflects marginal cost, while prices for inframarginal consumption are set so that the utility breaks even. Another strategy, Ramsey pricing, is to set different prices for different customers or customer groupings according to the magnitude of their elasticities of demand.

### Seasonal Rates

The simple emulation of marginal-cost pricing requires that all customers purchasing a good should be charged the same price, namely the good's

---

* The latter approach avoids the undesirable cross-subsidization implicit in Ramsey pricing by applying a uniform upward or downward scaling to marginal-cost-based prices so that revenue equals total cost.

marginal cost. But suppose that it costs more to provide the good during certain periods than others. Suppose, for instance, that a water agency does not have one but several sources of supply, and that its most expensive sources are required only when demand is high, whereas its cheaper sources are sufficient when demand is low. How then should the utility price its water? Under marginal-cost pricing, the price should reflect the cost of the last unit produced. But since this cost may differ over different production periods, we get different prices at different times. This can be considered an example of peak-load pricing: the service is priced higher during the peak demand period, when cost is high, than during the off-peak period, when cost is low. For some utilities the major variation in demand occurs during different parts of the day (e.g., early evening, middle of the day, night) or during different days of the week (e.g., weekday versus weekend). For water, while there are daily and weekly cycles, the main variation is seasonal—summer versus winter—because outdoor water use varies with seasonal changes in temperature, precipitation, and plant evapotranspiration. For this reason, the terms *seasonal pricing* and *seasonal rates* are commonly used in the water industry when discussing peak-load pricing. We next discuss both the theory and the application of peak-load pricing for water service in terms of economic efficiency, cost allocation, and incentives for conservation.

**Principles of Peak-Load Pricing.** It is safe to say that the economics profession is fairly unanimous in its prescriptions for pricing capacity required to meet peaks in demand. In general, there are two. Both follow from marginal-cost theory, and both assign capacity costs to those that are causally responsible for them. One is appropriate in instances where variation in demand is random, and therefore unpredictable; the other is appropriate when the variation in demand is systematic, and therefore predictable, at least to a reasonable degree of accuracy. As it turns out, both situations occur with urban water service. Fire service is perhaps the best example of an unpredictable demand variation that requires that excess capacity be on hand. Summer sprinkling is an example of a much more predictable demand variation that also requires that excess capacity be on hand. Before discussing in more detail each pricing prescription, however, additional insight into the nature of the service provided by a water utility may be useful.

A fundamental condition of urban water service is that the utility must be ready to deliver at any time. In this sense, as Hirshleifer et al. aptly state, the utility must stand "ready to enter into a contract for delivery at the option of the buyer." It is the customer rather than the utility that is in the proverbial driver's seat determining the quantity of service the utility must provide at each given moment. To oblige the demands of its customers with a given level of reliability, the utility requires an excess capacity above the average level of demand it can expect. All urban water systems are designed with this in mind.

Essentially, then, a customer holds an option to utilize this reserve capacity, and can exercise this option at any time. The cost of the option to the utility is the cost of providing the extra capacity necessary to meet maximum expected demand. Efficiency requires that the customer's willingness to pay for the option equal the cost of the option. In other words, the customer gaining access to additional capacity must be willing to pay the incremental cost of providing it. If the price for additional capacity is less than its marginal cost, over time too much capacity will be demanded. If customers were to face the true cost of the additional capacity, they would scale back their demand and apply the savings to more valued alternatives. Conversely, if the price is above its marginal cost, too little will be demanded. In this case, if customers were to face the true cost of additional capacity, they would scale back their demands for other goods and services and apply the savings to the purchase of more capacity.

With this in mind, Kahn states the economist's rule for pricing capacity as follows:

> If the same type of capacity serves all users, capacity costs as such should be levied only on utilization at the peak. Every purchase at that time makes its proportionate contribution in the long-run to the incidence of those capacity costs and should therefore have that responsibility reflected in its price. No part of those costs should be levied on off-peak users. (Kahn 1988)

This is the rule of "Peak Responsibility." Those that create the peak should pay for the peak. However, as we alluded to earlier, identifying those responsible depends on whether the capacity is required to satisfy random or systematic spikes in demand. If it is random and can occur at any time (e.g., fire service), the procedure is to allocate the cost of the necessary capacity according to the maximum instantaneous demand a customer or class of customers can be expected to place on the system (Hirshleifer, Dehaven, et al. 1960). This can be accomplished with a "capacity" charge for the extra capacity that a utility holds in readiness for its customers' unexpected demands. It is important to note that this should not be a volume charge but rather a fixed charge that reflects the cost of providing capacity sufficient to ensure a reasonable degree of reliability in service.

If, on the other hand, the variation in demand is systematic, occurring during a certain period or periods (e.g., sprinkling demand), then the price for service during the peak period(s) should include the cost of capacity that makes consumption at the peak level possible.* This results in a very different pricing formulation than when the variation in demand is random. In this case, a volume charge is not only appropriate, it is necessary to obtain efficiency. When peak demand occurs during a specific period, then any consumption during that period contributes to the peak and thus to the need for capacity. In other words, any consumption during the peak period is, in part, responsible for the capacity required to satisfy it. Whether a customer's consumption is small or large during that period is of no relevance, since it is the sum of all demands that creates the peak. Therefore, the price of a unit of water during the peak period for all customers should reflect the cost of providing this additional amount of water.

It is also true, however, that capacity available for service during the summer peak is also available during the winter off-peak. Indeed, if summer ceased to exist, part of the capacity would still be needed to serve winter. Should not winter, then, be responsible for the share of capacity it requires? Should summer, then, be charged only for the *extra* capacity above and beyond that needed to serve winter? The answer to both questions, in general, is no.[†] As Kahn states:

> Any attempt to shift capacity costs to the offpeak demands, by raising prices for that service above its own separate, incremental cost . . . , will cause available production capacity at that time . . . to be wasted, and would cut off purchasers willing to pay the additional cost of serving them. Any reduction of the peak . . . price below the full joint cost . . . would stimulate additional purchases at the peak, requiring additions to capacity that would not be made if buyers had to pay the full opportunity costs of the additional resources required to supply them. (Kahn 1988)

In the case of water service, it is the second instance that is of most importance. The underpricing of service during the peak period results in overconsumption, and, in the long run, will encourage overdevelopment of water resources. From a practical standpoint, would adjusting the peak-

---

* Note that the timing of the peak is different for water than, say, electricity. With the latter, the peak tends to occur at certain hours of the day (e.g., late afternoon) and on weekdays instead of weekends. While daily and hourly variation in water use also occurs, the ability to store large amounts of water and the need for fire-flow capacity make it much less significance than for electricity. For water, the seasonal peak is the focal point of peak pricing issues.

† Actually, a better general answer to the two questions posed would be "it depends." However, answering the questions in this way would lead us too far astray. The reader interested in a very lucid and detailed discussion of capacity costing from the economist's point of view is referred to Hirshleifer, Dehaven, et al. (1960) and Kahn (1988).

period price to reflect the full cost of water service have much impact on consumption? The answer depends on the responsiveness of demand to price. The empirical evidence suggests that residential demand for water service during the summer (the peak period for water service) is markedly more elastic than for the winter, and water demand for outdoor uses is more elastic still. Since outdoor use tends to drive the peak, this would suggest that underpricing service during the peak period could indeed have a significant impact on consumption.

**Implementing Seasonal Rates.**   Before a utility can implement seasonal rates, however, it must address several additional considerations. These include the stability of the peak period with respect to price and time, measurement, and administrative feasibility. We address each of these in turn.

A legitimate concern for a utility is that the institution of peak-load pricing may simply shift the peak to another period. Faced with a higher price in the peak period, customers shift, en mass, to the off-peak period where price is lower.* Whether this is a significant possibility depends on the cross-price, elasticity between the two periods (i.e., the degree to which a change in price in one period affects the demand for the good in another period). In the case of seasonal rates for water, it is very unlikely that this would pose a problem. To a great extent, the difference in demand between the peak and off-peak periods is determined by factors other than price, such as climate. While an increase in the price of water during the peak period is likely to induce a decrease in consumption during that period, it is not likely to induce an appreciable increase in consumption during the off-peak period. Overall, the level of demand is likely to fall.

Another equally counterproductive possibility exists, however. Suppose that upon instituting a seasonal rate schedule the utility discovers that while the average level of demand during the peak period has declined, the actual peak day or hour demand, which determines the capacity requirement, has remained unchanged. This phenomenon has been labeled "needle peaking" (Beecher, Mann, et al. 1990). If we were to graph the pattern of consumption the peak would appear as a needle. Overall, the demand for water during the peak period has declined, which is good from the point of view of conservation, but the level of capacity required to meet demand has remained unchanged, which may be very bad from the point of view of utility finance. It means that a greater amount of capacity must

---

* There are many examples, such as telecommunications (where customers switch some calls and faxes to the evening), airlines (where people switch vacations to off-season months), and electricity (where, for example, electric storage heaters are designed to use electricity during the evening hours).

stand idle for a longer period of time, thus deteriorating annual load factors and possibly eroding revenue.* Whether this is a significant risk is largely an empirical question that must be answered with the hard hand of experience.

A final consideration is that the peak may shift through time for reasons unrelated to price. Changes in technology, preference, or social policy each have the possibility of shifting the peak period. The relevance of this is that consumers may base long-term investments in water conservation, in part, on the cost of water during the peak period, which ideally would reflect the long-run marginal cost of additional supply. As the peak shifts, so too will the long-run marginal cost, and thus what was a wise investment before the shift may no longer be so wise after the shift. Again, this does not appear to be a considerable risk in the case of water service, where hydrologic conditions maintain a very stable peak.[†]

Before a utility can institute peak pricing it must have some way to differentiate between peak and non-peak consumption and to measure system costs that are associated with peak-period consumption versus those that are associated with off-peak consumption. In the case of urban water service, identifying the peak season of demand is straightforward. Low winter use and high summer use are fairly typical for urban water agencies in the west. Urban water utilities also experience daily peaks and valleys in demand in a manner similar to those seen in the electricity supply industry. To measure these demands, however, requires meters that can mark both the amount of consumption and the timing of consumption, which currently is beyond the grasp of almost all water utilities. Therefore, the practical application of peak-load pricing for retail water service is restricted to monthly demands.

Measuring system costs associated with peak consumption often is less straightforward, though not impossible. Marginal costs associated with a particular service or class of customer should be assigned accordingly. This corresponds to the pricing rule that a good must cover its variable operating cost. Many system costs, however, are joint costs, and it is sometimes difficult to determine which use or combination of uses is the driving factor. This is particularly true for assigning causal responsibility for

---

* However, if the decline in demand is purely price motivated, and demand is price inelastic (as empirical studies suggest), then a positive change in utility revenue would result from a price increase.

[†] Technology, too, has been shown to exert its influence on water consumption patterns, sometimes in rather amusing ways. For example, starting in the late 1940s and early 50s, water agencies started to detect small evening peaks in demand recurring at fifteen- and thirty-minute intervals. Eventually, it was determined that these peaks were the result of concentrated use of bathroom fixtures during TV commercial breaks. Perhaps with the adoption of ultra-low-flush toilets, evening demand will again resemble its pre-TV pattern.

capacity. Fire service and peak summer service are the two principal candidates. As we discussed earlier, the pricing rules for the two types of service are very different—one being a fixed charge, the other being a volume charge. Thus, it is important that one not be assigned cost responsibility for the other. In general, fire-flow requirements will determine sizing for the distribution and storage network in the immediate vicinity of customers, while seasonal peak-day demand will determine sizing for most transmission lines, pumping stations, and treatment plants.

Finally, before adopting a seasonal rate structure—or any price structure for that matter—a utility should be reasonably confident that the gains from doing so will exceed the costs. Marginal cost principles, in addition to guiding the pricing of a good such as water, also can be applied to the feasibility and cost-effectiveness of a particular price structure itself. It is always possible that the cost of designing and administering a peak-load price structure will exceed its benefits. Whether this is a likely occurrence depends on numerous factors, including: the cost differential implicit in serving peak and off-peak uses; the price responsiveness of demand during the peak; the cost of measuring and assigning system costs to peak and off-peak uses; and the cost of metering and billing peak and off-peak uses. For example, given the existing stock of metering equipment in place, it is very unlikely that potential water savings and load factor improvement derived from time-of-day rates would exceed the added cost of metering and billing necessary to implement them. On the other hand, it is very possible that seasonal pricing, which could utilize existing metering devices and would not require drastic changes in billing procedure, would be efficient, though not always guaranteed.

**Seasonal Rates Summary.**   Seasonal rates are a potentially effective means for realizing more efficient use of scarce water resources when demands on a water utility's system vary systematically across seasons. Their primary advantage is that they provide to consumers an accurate signal of the cost of consumption, including the cost of capacity, in a given period. In this regard, seasonal rates have several advantages over more traditional approaches to pricing capacity, including:

- Consumption within periods responsible for capacity costs pay for those costs. Traditional approaches typically spread these costs over all periods. This can increase inefficiency and decrease equity. It increases inefficiency by underpricing water service in the peak period, thus encouraging too much consumption, and by overpricing water service in the off-peak period, thus encouraging too little consumption. It decreases equity because off-peak users, by paying a share of cost that they are not causally responsible for, implicitly subsidize the consumption of peak users.

- All uses during the peak period are recognized as contributing to and are charged for the cost of meeting the peak.

- Ideally, seasonal rates will also reflect the full cost of capacity required to meet the peak rather than just that portion in excess of average demand. Traditional approaches, on the other hand, go to great lengths to identify whether capacity is meeting average demand or peak demand requirements. In fact, the capacity is jointly meeting both, and causal responsibility for costs depends on the relative magnitudes of peak and off-peak demands. In cases where the differential between the two demands is large, the peak period will bear responsibility for the costs.*

The practical considerations of developing and applying a seasonal rate also must be taken into account. Seasonal rates may present a somewhat more complicated, and hence more expensive pricing formulation than one that does not use different rates for different periods. Therefore, the costs of design and administration must be carefully weighed against the potential gains in efficiency. Still, the increasingly common occurrence of a seasonal rate differential would suggest that these costs are not prohibitive for many urban water utilities.

## Increasing Block Rates

Increasing block rates are rates where the volume charge increases for each successive quantum of water. For example, a rate that charges $1 per unit for the first 100 units and $1.50 per unit for all subsequent units is an increasing block rate. Many utilities now employ increasing-block rates of one variety or another.

An increasing-block rate is often proposed to meet two objectives: (1) to promote customer conservation; and (2) to satisfy revenue sufficiency constraints. Its potential to promote customer conservation is well described by the following passage:

> The impact [on conservation] of an increasing block rate design is best illustrated by comparing it to the simplest alternatives uniform design where all water is sold at the same price. Because of revenue sufficiency and other constraints, it can be assumed that both rates are initially designed to recover the same total revenue. The increasing block design, then, must contain one or more prices which are higher than the uniform design, and one or more which are lower. Customers facing the higher prices at the margin will, in theory, use less water than they would under the uniform design; customers facing lower prices at

---

* The exception being the capacity required for fire-flow.

the margin will use more. The increasing block design will conserve water if the sum of decreases in use exceeds the sum of increases. (Metropolitan Water District of Southern California 1991)

The expectation is that demand in the high blocks will be more elastic than demand in the low blocks, resulting in a net decrease in water use. The empirical evidence on elasticity suggests that this is not an unreasonable expectation; but neither is it a guaranteed result. The schedule's design, customer attributes, and regional climate will be important factors to any outcome.

Increasing-block rates also are advanced as a means to preserve revenue neutrality. For example, when marginal costs are increasing, a marginal-cost-based rate will collect more than the utility's revenue requirement. A commonly proposed solution to this is an increasing-block rate that sets the marginal price for just the last block and sets the price for prior blocks so that the utility breaks even. If all customers face the marginal cost for at least some of their consumption, the efficiency properties of a marginal-cost-rate design will be preserved. The problem, of course, is that it is exceedingly difficult to design a rate schedule that results in every customer facing the marginal rate for at least some of their consumption. More typical is a rate where some customers pay a high price and some pay a low price. Unless this is cost-justified, questions of equity will arise.

**Case Studies of Increasing Block Rates.** We noted previously that increasing-block rates have been used by a number of water agencies at various times. In 1990, for example, 16 of the 60 plus water agencies served by The Metropolitan Water District of Southern California had increasing-block rates, as shown in Table 5-1 (some other agencies within Metropolitan's service area have adopted increasing-block rates since then, including Los Angeles whose experience is discussed in this section). Those rates were not necessarily adopted as conservation rates, but it may be instructive to analyze them from that perspective. In most cases, there are only two blocks, as opposed to having many, finely graduated blocks. There is considerable variety with regard to the magnitude of the price differentials between the blocks: in nine cases, the differentials are in the range of 5 to 15 percent; in three cases, they are in the range of 35 to 65 percent; and in five cases, they exceed 100 percent. What is most striking is the location of the switch points. In 10 of the 16 rates, the switch point is at 125–250 gallons per account per day. One thinks of a typical single-family residential account as having three or four people in the household with each having an indoor use of between 60–80 gallons per day. Thus the total indoor use for the typical account may range between 210–280 gallons per day. Added to this amount is outdoor use, which will vary considerably by

**Table 5-1.** Retail Agencies in MWD Service Area
with Block Rates in 1990

| Agency | Population served | Rate structure (marginal price in $/AF) (g/d—gallons/day/account) | |
|---|---|---|---|
| Los Angeles County | | | |
| Beverly Hills | 34,300 | $0 | up to 250 g/d |
| | | $433 | up to 374 g/d |
| | | $469 | up to 748 g/d |
| | | $503 | up to 1500 g/d |
| | | $525 | above 1500 g/d |
| Glendale | 166,100 | $159 | up to 125 g/d |
| | | $182 | above 125 g/d |
| Inglewood | 103,000 | $601 | up to 199 g/d |
| | | $806 | above 199 g/d |
| Las Virgenes | 54,400 | Zone 1: | |
| | | $283 | up to 150 g/d |
| | | $828 | up to 300 g/d |
| | | $558 | above 300 g/d |
| | | Zone 2: | |
| | | $923 | up to 150 g/d |
| | | $1032 | up to 300 g/d |
| | | $1198 | above 300 g/d |
| Long Beach | 419,800 | $177 | up to 125 g/d |
| | | $362 | above 125 g/d |
| Pasadena | 132,200 | $148 | up to 374 g/d |
| | | $317 | above 374 g/d |
| Pomona | 118,000 | $136 | up to 150 g/d |
| | | $227 | above 150 g/d |
| Santa Monica | 96,500 | $266 | up to 250 g/d |
| | | $301 | above 250 g/d |
| Orange County | | | |
| Fountain Valley | 55,600 | $331 | up to 250 g/d |
| | | $392 | above 250 g/d |
| San Clemente | 35,100 | $0 | up to 125 g/d |
| | | $540 | up to 648 g/d |
| | | $810 | above 648 g/d |
| Riverside County | | | |
| Riverside | 199,400 | $65 | up to 150 g/d |
| | | $261 | above 150 g/d |
| San Diego County | | | |
| El Cajon | 83,200 | $365 | up to 250 g/d |
| | | $421 | above 250 g/d |

**Table 5-1.** Retail Agencies in MWD Service Area
with Block Rates in 1990 (*Continued*)

| Agency | Population served | Rate structure (marginal price in $/AF) (g/d—gallons/day/account) | |
|---|---|---|---|
| La Mesa | 52,200 | $365 | up to 250 g/d |
| | | $421 | above 250 g/d |
| San Diego | 1,027,360 | $409 | up to 125 g/d |
| | | $468 | above 125 g/d |
| Ventura County | | | |
| Oxnard | 124,000 | $0 | up to 125 g/d |
| | | $457 | up to 250 g/d |
| | | $482 | above 250 g/d |
| Simi Valley | 94,500 | $283 | up to 685 g/d |
| | | $414 | above 685 g/d |

SOURCE: Metropolitan Water District of Southern California, Water Conservation Pricing, Approaches of The Metropolitan Water District, Staff Report, October 1990.

account, but that could easily average an additional 100–200 gallons per day, bringing the total to between 310–480. What these numbers suggest is that the majority of the utilities—10 of 16—employing increasing-block rates in Metropolitan's service area are locating switch points at levels too low for most single-family households to attain without a very substantial change in the pattern of household water use. A more appropriate location to induce households to switch from above the switch point to below might be somewhere in the range of 300–500 gallons per day.

There are some cases where water districts have adopted increasing-block rates with the promotion of conservation as an explicit consideration in the design of the rate structure. We consider three of the more informative examples. The first involves a small water district serving irrigators in the San Joaquin Valley rather than M&I users. While the agricultural context is very different from what we have been discussing, in this instance, the basic principles of rate design remain the same and are particularly well illustrated by this example. The second example is Tucson Water, which has employed increasing-block rates, in part to promote conservation, for well over a decade. The third is the LADWP, which adopted increasing-block rates in 1993.

What these three utilities have in common is that they attempted to use the rate structure to change the distribution of water use by targeting consumption at the high end of the range of consumption. By way of explanation, when one sees data on distribution of water usage by individual residential accounts there is almost always a distinctive long right tail—a small fraction of the users at the high end account for quite a substantial fraction of total use. Rather than the symmetric, bell-shaped distribution of the normal probability distribution, one finds something closer to the distribution in Figure 5-2. For instance, the top 24 percent of single-family residential accounts in the LADWP service area in 1988 accounted for 47 percent of the total use, and the top 10 percent accounted for almost 27 percent of the total use. Overall, usage per residential account varied from as little as 25 gallons per day to as high as 22,400 gallons per day (there were five accounts using this amount). Some of this variation is due to differences in household size or lot size—but not all. Some of it is due to lifestyle, habit, and preference (e.g., whether people bother to fix leaky faucets, how much attention they pay when watering the yard, etc.).*

The existence of this distinctive pattern of household water use, with a long right tail, raises a fundamental question about the design of conservation-oriented water rates. If one wants to reduce overall water use by $x$ percent, should one aim to shift the entire distribution of water use to the left by $x$ percent, or should one seek to change the shape of the distribution, pulling in the right tail in such a way as to reduce overall use by $x$ percent. The first strategy aims to change every customer's water use. The second targets customers with substantially above-average consumption; it rests on the notion that, if one could just attract their attention and induce them to change their pattern of use, this could lower total use without having to impact the behavior of the entire customer base. Which is better is an empirical matter that will certainly vary with circumstances. If targeting a large incentive at the fraction of users in the right tail is more effective than offering a smaller incentive for everybody, this strengthens the case for the strategy of changing the shape of the distribution. That was the objective in each of the three case studies, to which we now turn.

*Example 1: Broadview Water District.*   Broadview water district is located near Firebaugh, California. It serves an area of about 10,000 acres of farmland, making it one of the smaller agricultural water districts in the state. Broadview is one of several irrigation districts on the west side of the San Joaquin Valley implicated in the pollution of Kesterson Wildlife Refuge and the nearby San Joaquin River with selenium-laden drainage water. Interim water quality standards for reaches of the San Joaquin River that serve as outfall for Broadview drainage were announced in 1987–88. To comply with the standards, Broadview would have to reduce its drainage into the river by about 15 percent. It was determined that this could prob-

---

* The variation in use does *not* reflect difference in price, since these households were all paying the same flat-rate price in 1988.

ably be done by reducing the amount of irrigation water applied to crops by approximately 10 percent, since overirrigation was identified as a major cause of high drain flows.

In October 1988, the start of the 1989 crop year, the district announced a new rate structure. Previously, the district had charged a uniform commodity charge of $16 per acre-foot of delivered water, together with a fixed assessment of $42 per acre served by the district. Part of these charges covered the cost of the districtwide drainage collection and conveyance system. The annual cost to operate and maintain this system amounted to about $21 per acre-foot of collected drainage water or, when averaged over the roughly 25,000 acre-feet of water delivered annually by the district, about $3.08 per acre-foot. The district felt, however, that a surcharge of $3.08 per acre-foot would be too insubstantial to elicit a significant reduction in on-farm water use. Instead, it adopted a two-tier rate structure; water in the first block continued to be priced at $16 per acre-foot; water in the second block would be priced at $40 per acre-foot. The idea was to use the revenues from the sales in the second block to cover the drainage system costs. Separate switchpoints were set for each crop grown in the district; these were set at 90 percent of the districtwide average water usage for the crops over the period 1986 through 1988, as shown in Table 5-2.* If irrigators behaved toward the incentive as anticipated, applied water within the district would decline by about 10 percent as desired.

The results of the new rates in 1989 and 1990 are shown in the last two columns of Table 5-2. In four cases, there was already some reduction in water use in the 1989 growing season (a drought year), but not for cotton and wheat. With those two crops, efforts at conservation were offset by unusually hot temperatures during key parts of the 1989 growing season. For cotton and wheat, the most pronounced responses to the new rates occurred in 1990. Cotton is the most important crop grown in the district, planted on about half the district acreage. Field-level data for cotton indicate the effect the rate had on water use decision, as shown in Table 5-2. Cotton was grown on 32 fields in 1989 and 33 fields in 1990. In 1989, the modal application rate (i.e., the rate that was applied on the greatest number of fields) was 3.6 acre-feet per acre. On eight fields, the application was higher than this amount. In 1990, the modal application rate fell to 3 acre-feet per acre, very close to the switchpoint of 2.9 acre-feet per acre, and only one field had an application rate above the 1989 modal rate. Since 1990, the trend in water use for both cotton and other crops has continued

---

* This corresponds to the notion that switchpoints, to be effective, must be set at levels that are within the reach of customers being asked to conserve. If the district had simply set a single switchpoint for all crops, irrespective of individual crop water requirements, the effectiveness of the rate would have been greatly reduced.

**Table 5-2.** Crop-Specific Water Use in Broadview Water District, 1986–1990

| Crop | Average use 1986–1988 (af/acre) | Switchpoint (af/acre) | Average use 1989 (af/acre) | Average use 1990 (af/acre) | % Change 1986–88 to 1990 |
|---|---|---|---|---|---|
| Cotton | 3.20 | 2.9 | 3.34 | 2.84 | –11 |
| Tomatoes | 3.22 | 2.9 | 2.73 | 3.03 | –6 |
| Melons | 2.11 | 1.9 | 1.93 | 1.79 | –15 |
| Wheat | 2.30 | 1.9 | 3.02 | 2.18 | –5 |
| Sugarbeets | 4.58 | 3.9 | 3.73 | 2.54 | –45 |
| Alfalfa seed | 2.06 | 1.9 | 1.84 | 1.88 | –9 |

SOURCE: Dennis Wichelns and David Cone. "Irrigation district programs motivate farmers to improve water management and reduce drain volume." Presented at U.S. Committee on Irrigation and Drainage, 12th Technical Conference on Irrigation, Drainage and Flood Control, San Francisco, CA, November 13–16, 1991.

downward, though the drought has been a confounding factor that makes it difficult to determine the extent to which the new rates have encouraged this trend. It is clear, nevertheless, that there has been a substantial reduction in water use in Broadview and that this is likely to be permanent. Numerous improvements in irrigation practice since 1989 have now become fixtures in the district. There is little doubt that these changes were greatly encouraged by the new rate structure.

There are several features of this story, some not yet mentioned, that act to make it unique, but also highly illustrative of the factors that contribute to the effectiveness of a conservation rate structure. First, the district was able to tailor its block rates in a way that enabled it to motivate all of its customers. Rather than apply a single rate to a heterogeneous group of users, it differentiated rates by crop, and thus was able to apply individual rates to more or less homogeneous groups of users. This allowed it to tailor the price signal to increase its effectiveness. Second, farmers within the district had detailed knowledge of their water use, monitored it fairly closely, and faced relatively drastic financial consequences if their water use was excessive. In other words, the farmers were very cognizant of the relationship between price and quantity, which is fundamental to customer responsiveness to price. Third, farmers in the district had direct access to information on possible conservation strategies, knew how well their neighbors were doing, and understood the normative expectations of the district. Indeed, water usage rates were posted by the district each month, and anyone deviating from the norm could expect a good razzing from his fellow growers. Thus, the rate setting environment satisfied each of the factors discussed earlier that influence price responsiveness and make conservation rates effective: rates were tailored to target homogeneous groups of users; the rate differentials presented users with significant incentive to curtail use; users themselves had a firm understanding of their use and how they could curb it; and normative expectations were clearly established and adhered to.

In an urban water district the situation is different in each respect. Urban utilities seldom face a homogeneous group of users for which to tailor rates. Typically, a utility's customers are differentiated only up to the point of primary class (e.g., residential, commercial, industrial, and institutional). Within any class, however, the degree of customer heterogeneity is still apt to be very great, which will make it much more difficult to develop a rate with price differentials and switchpoints relevant to a wide variety of customers. It is also unrealistic to expect an urban utility's customers, particularly residential customers, to understand and be aware of their water use to the same degree as the farmers of Broadview. While water is central to our day-to-day tasks, we do not face the same degree of risk to income if we use a little more than we should or fail to conserve as much as we should, and thus it is not too surprising that we are fairly ignorant of our water uses. For example, when asked how much water they consumed in a day, surveyed customers in Southern California were frequently off by a factor of ten. Thus, a key factor contributing to the effectiveness of conservation rates—knowledge of use—may be missing or muted in the urban setting. Finally, normative expectations become increasingly difficult to convey as the size of the group increases. In Broadview, the small number of growers made it relatively easy to impose a group standard. The same will seldom be the case for an urban utility.

Given these differences, is it reasonable to expect the same type of conservation potential from an inverted block rate in an urban setting? Probably not to the same degree as experienced in Broadview, but the experiences of at least one utility—Tucson Water—indicate that an inverted block rate can measurably decrease high-end consumption and result in significant water savings.

*Example 2: Tucson Water Department.* Tucson Water first instituted an inclining block rate in the mid-1970s. In 1977, it combined a four-block rate for summer months with a flat rate for winter months. By 1986, it had switched to a six-block rate for both winter and summer, though rates in the summer were set higher than in the winter. Water rates were increased each year between 1977 and 1986, though in inflation-adjusted dollars water bills either remained constant or declined until 1982. Starting in 1982, rates were adjusted in a way to make it more expensive in real as well as in nominal dollars to consume above-average amounts of water, but left water bills for customers using average or below-average amounts mostly unaffected. The notion was to motivate without coercing customers using considerably above-average amounts of water to scale back their consumption.

Between 1982 and 1986, water bills calculated with inflation-adjusted dollars for single-family residential customers with usage three times 1978–79 average usage increased by 26 percent. Over the same period, bills for customers with usage equal to 1978–79 average usage increased about 7 percent, while bills for customers with usage equal to one-half this amount increased about 9 percent. The impact of this rate adjustment on consumption, particularly with respect to usage in the upper blocks, was analyzed by Cutherbert and Nichols (1987). Their analysis indicates that the rates were relatively successful in shifting consumption out of the upper blocks. Normalizing for weather and other factors, they found that for the period 1982 to 1986, annual usage in the upper three blocks as a share of total residential usage declined from 8 percent to 6.6 percent. More important, however, weather-normalized average monthly usage in the peak usage months of June–July was 11 percent lower in 1986 than in 1982, but was left unaffected during the nonpeak winter months, clearly suggesting that the rate structure was effectively curtailing discretionary outdoor use during the peak season when the utility is most at risk for shortage. Annual water savings were estimated to be 550,000 CCF in 1983, 1,000,000 CCF in 1984, 1,500,000 CCF in 1985, and 1,300,000 CCF in 1986. By 1986, then, estimated annual savings from shifting the distribution of consumption away from the high-end range equaled an amount of water sufficient to supply more than 9000 customers.

*Example 3: Los Angeles Department of Water and Power.* In the summer of 1991, in the face of a serious drought that had required rationing and emergency water rates to cope with a 15 percent shortfall in supply, Mayor Bradley appointed a Blue Ribbon Committee to consider LADWP's water rates for the future. The committee made an extensive analysis of LADWP's costs and then proposed an increasing-block-rate structure designed specifically to target higher-use customers.* The committee was committed to the basic principle of marginal-cost pricing, but wanted to ensure revenue sufficiency in the face of a rising marginal cost of new supply sources. It saw the increasing-block rate structure coupled with carefully designed automatic rate adjustment mechanisms as a means of providing revenue sufficiency while at the same time promoting conservation through a targeted incentive. The committee also was concerned that the adoption of the new rate structure might be misconstrued by some simply as a revenue-generating device rather than as a means to promote more efficient use of water. Therefore, the structure was designed to be revenue neutral in com-

---

* One of the present authors (MH) served as a technical adviser to the committee.

parison to its predecessor for the first year following implementation.* Another advantage of the two-tier structure was that it protected more conserving users, whom the committee felt should be rewarded for their efforts, not penalized. After some analysis, the committee decided that reclaimed water should be taken as the marginal source of supply for the purpose of estimating off-peak marginal cost.[†] In addition, it proposed seasonally varying rates, based on the change in marginal cost between off-peak and peak periods, with the peak users covering certain of the capacity costs of treatment, transmission, and distribution.[‡] The committee felt that the incentive offered by two-tier rates could still be effective for consumers who were below the switchpoint, as long as it was sufficiently close that the higher price loomed in their consciousness and could influence their behavior. Finally, the committee felt strongly that there should be the same rate structure throughout the city, as opposed to having different structures (e.g., different switch points or different prices) in different areas.

The rates that the Blue Ribbon Committee recommended are shown in Tables 5-3 and 5-4. The rates are structured differently for single-family residences compared to other users. For single-family residences, the switchpoint is located at 525 gallons/account/day. The other customer classes—multifamily residential, commercial, and industrial—are considerably more heterogeneous than the single families and it was felt that, for them, the switchpoint should be based not on some absolute level of use that would be the same for all users within the class but rather on a *relative* level of use, namely usage in the winter. Thus, in the winter there is a single block rate; in the summer, the first block applies to consumption up to 125 percent of winter consumption, and the second-block rate applies to consumption beyond this level.[§] The second-block rate is the same for *all* users and is based on the estimate of LADWP's marginal cost of supply; it differs between summer and winter because of the peak-load pricing

---

* After the first year, revenue would be allowed to vary from that generated under the old rate schedule. The rate ordinance guarantees LADWP a base revenue of $277 million and includes a Water Revenue Adjustment Factor to adjust rates if revenues fall short of targets.

[†] It was understood that this could change over time as new developments occurred, such as increased supplies becoming available from demand-side management or water markets sales by agricultural users.

[‡] Having adopted flat rates in place of declining-block rates in the previous drought (1977), LADWP introduced seasonally differentiated flat rates in 1985. The summer rate was initially set at 15 percent higher than the winter rate; by 1992, the differential had risen to 25 percent. The committee felt, however, that this differential was too small to reflect the real differences in correctly calculated peak and off-peak marginal costs, as well as too small to attract much notice from water users.

[§] It should be noted that sewer charges in the LADWP service area are based on the volume of water used during the winter period. This provides a complementary incentive against artificially boosting wintertime use for the sake of lowering summer water bills.

**Table 5-3.** Water Rates Proposed by LADWP
Blue Ribbon Committee

| | | Normal year rates | | |
|---|---|---|---|---|
| | Price in low block ($/CCF) | Switchpoint | Price in high block ($/CCF) | |
| | | | Winter | Summer |
| Residential | | | | |
| Single-family | $1.71 | 525 gallons/day | $2.27 | $2.92 |
| Multifamily | $1.71 | 125% of winter use | NA | $2.92 |
| Nonresidential | $1.78 | 125% of winter use | NA | $2.92 |
| | | Drought year rates | | |
| | Price in low block ($/CCF) | Switchpoint | Price in high block ($/CCF) | |
| Residential | | | | |
| 10% Shortage | $1.71 | 475 gallons/day | $3.70 | |
| 15% Shortage | $1.71 | 450 gallons/day | $4.44 | |
| 20% Shortage | $1.71 | 425 gallons/day | $5.18 | |
| 25% Shortage | $1.71 | 400 gallons/day | $6.05 | |
| Multifamily | | | | |
| 10% Shortage | $1.71 | 115% of adjusted winter use | $3.70 | |
| 15% Shortage | $1.71 | 115% of adjusted winter use | $4.44 | |
| 20% Shortage | $1.71 | 110% of adjusted winter use | $5.18 | |
| 25% Shortage | $1.71 | 110% of adjusted winter use | $6.05 | |
| Nonresidential | | | | |
| 10% Shortage | $1.78 | 115% of adjusted winter use | $3.70 | |
| 15% Shortage | $1.78 | 115% of adjusted winter use | $4.44 | |
| 20% Shortage | $1.78 | 110% of adjusted winter use | $5.18 | |
| 25% Shortage | $1.70 | 110% of adjusted winter use | $6.05 | |

**Table 5-4.** Normal Year Rates Adopted by L.A. City Council

| | | Normal year rates | | |
|---|---|---|---|---|
| | Price in low block ($/CCF) | Switchpoint | Price in high block ($/CCF) | |
| | | | Winter | Summer |
| Residential | | | | |
| Single-family | $1.14 | 575 gallons/day | $2.23 | |
| | | 725 gallons/day | | $2.98 |
| Multifamily | $1.14 | 125% of winter use | NA | $2.92 |
| Nonresidential | $1.21 | 125% of winter use | NA | $2.98 |

design. The rate for the first block varies among customer classes and is set so as to meet the revenue targets for that class.

In addition, there is a separate set of rates for drought years. The idea is to set down ahead of time the principles that will be followed when it comes to adjusting water rates in the course of a drought. The same type of block-rate structure still is applied in a drought year, but it is modified in two ways to adjust to the shortage. First, the switchpoint that the second block commences is reduced, roughly in proportion to the severity of the shortfall. This means that the higher price will be triggered sooner during a drought than during normal supply years. Second, the rate charged in this second block is raised to equal what the committee estimated to be the rationing price that would equilibrate demand to supply, given the shortfall.*

In January 1993, the L.A. City Council adopted a rate ordinance which closely followed the committee's recommendations. The main change was to raise the switchpoint for single-family residential accounts and differentiate it by season, placing it at 575 gallons/account/day in the winter and 725 gallons/account/day in summer. This was done mainly to accommodate the interests of residents living in the San Fernando Valley who face a warmer climate and tend to have larger lots than residents in the downtown and coastal areas.

LADWP staff had estimated that 71 percent of the single-family residences would have a lower water bill in a normal year with the new rate structure it replaced. They could have a higher bill in some of the summer months when their usage spilled over into the higher priced block, but not of the year as a whole. Residents of the San Fernando Valley would be somewhat more adversely affected—only 61.4 percent would have a lower bill over the course of the year. By contrast, in the downtown and on the west side, respectively, 84 percent and 91 percent of the single-family accounts would have a lower bill with the new rate structure.

In fact, these predictions have borne out well during the first 12 months with the rate experiment. By the end of the fall of 1993, however, it had run into strong political opposition from the San Fernando Valley. Several factors were responsible. In addition to the change in the water bill, there had been an increase in sewer charges, which are included on the water bill along with various other city charges. Customers attributed all of the change to the water rates. This coincided with the first hot months, September and October, when many households experienced a higher bill for the first time, after several months of lower bills. Moreover, a new

---

* These equilibrium prices were based on an analysis of LADWP's experience in the summer of 1991 when increasing-block rates with punitive upper rates were introduced temporarily to cope with the drought.

mayor of Los Angeles had been elected with considerable support from the San Fernando Valley. As a result, the Mayor reconvened the Blue Ribbon Committee November 1993, and directed it to reconsider the rate structure to make it more equitable for residents of the San Fernando Valley. The new commitee decided to preserve the two-block rate structure and the rates associated with the block, but to make the switchpoint vary among different groups of residential users in a manner that would fit their individual circumstances and be more equitable. The committee decided that users should be grouped according to climate and lot size. It identified three separate climate zones with the LADWP above 85°F. It identified four ranges of lot size: below 7500 square feet; 7500–10,999; 11,000–17,499; and above 17,499. Because each group was now more homogeneous, the switchpoint was set at about 120% of the existing median use of customers within the group, with some minor adjustment in the lowest groups. The resulting switchpoints are shown in Table 5-5. This change had two advantages: compared to the earlier rate design with a uniform switchpoint for all residential users, the new structure increased the number of users actually facing the marginal cost incentive to conserve, and it distributed the benefits and costs of rate reform more evenly among all of LADWP's residential customers. These rates were adopted by the Los Angeles City Council with some small increases in the switchpoint and the subdivision of the largest lot size category into two categories, to provide a more generous allowance for very large lots, as shown in Table 5-6.

**Table 5-5.** Switchpoints Recommended by the 1996 Blue Ribbon Committee

| Lot size (sq. ft.) | Summer average daily high | Usage charges at low initial block rate (gallons/day) | |
| --- | --- | --- | --- |
| | | Winter | Summer |
| <7500 | <75° | 325 | 400 |
| | 75–85° | 325 | 425 |
| | >85° | 325 | 425 |
| 7500–10,999 | <75° | 400 | 575 |
| | 75–85° | 400 | 625 |
| | >85° | 400 | 650 |
| 11,000–17,499 | <75° | 575 | 900 |
| | 75–85° | 600 | 975 |
| | >85° | 600 | 1000 |
| >17,499 | <75° | 725 | 1125 |
| | 75–85° | 750 | 1200 |
| | >85° | 750 | 1225 |

**Table 5-6.** Switchpoints Adopted by City

| Lot size (sq. ft.) | Temperature zone | Usage billed at low initial block rate (gallons/day) | |
| --- | --- | --- | --- |
| | | Winter | Summer |
| <7500 | Cool | 325 | 400 |
| | Moderate | 350 | 450 |
| | Hot | 350 | 475 |
| 7500–10,999 | Cool | 400 | 575 |
| | Moderate | 425 | 650 |
| | Hot | 425 | 675 |
| 11,000–17,499 | Cool | 600 | 900 |
| | Moderate | 625 | 1000 |
| | Hot | 625 | 1050 |
| 17,500–43,559 | Cool | 700 | 1125 |
| | Moderate | 725 | 1275 |
| | Hot | 725 | 1325 |
| >1 Acre | Cool | 900 | 1375 |
| | Moderate | 950 | 1550 |
| | Hot | 950 | 1625 |

**Increasing Block Rates Summary.**    An increasing-block rate is often proposed to promote customer conservation through higher rates without violating revenue sufficiency constraints. For example, increasing-block rates have been used by several utilities to reshape the distribution of consumption by discouraging high-end uses. They can be formulated based on average or marginal-cost data. If designed so that every customer pays the marginal cost of service for at least some of their consumption, they can mimic the efficiency properties of marginal-cost pricing. The heterogeneity of customer demands, however, makes this exceedingly difficult to accomplish. More typical is a block-rate structure that results in some people paying a higher price for service than others. Unless there are cost-based reasons to do so, concerns about equity will emerge.

# References

Beecher, J. A., P. C. Mann, et al. 1990. *Cost Allocation and Rate Design for Water Utilities.* Columbus, OH: The National Regulatory Research Institute.

Brown, S. and D. Sibley. 1986. *The Theory of Public Utility Pricing.* Cambridge, MA: Cambridge University Press.

Cuthbert, R. W. 1989. "Effectiveness of Conservation-Oriented Water Rates in Tuscon," *American Water Works Association Journal*, 81(33): 67–69.

Hall, Darwin C. and W. Michael Hanemann. 1996. "Urban Water Rate Design Based on Marginal Cost," in Darwin C. Hall (ed.), *Advances in the Economics of Environmental Resources Volume I: Marginal Cost Rate Design and Wholesale Water Markets*. Greenwich, CT: JAI Press Inc.

Hirshleifer, J., J. C. Dehaven, and J. W. Milliman. 1960. *Water Supply: Economics, Technology, and Policy*. Chicago, IL: University of Chicago Press.

Kahn, A. 1988. *The Economics of Regulation: Principles and Institutions*. Cambridge, MA: The MIT Press.

Metropolitan Water District of Southern California. 1991. Water Conservation Pricing Approaches of the Metropolitan Water District.

# 6

# Forms and Functions of Water Pricing: An Overview

Charles W. Howe
*University of Colorado*

## What Is the Price of Water?

The everyday notion of price is simple and straightforward: the price of a gallon of gasoline; the price of a restaurant dinner; the price of a particular car. In the water resources area, the concept of price is often more complex. Consider the following situation. The Northern Colorado Water Conservancy District distributes about 250,000 acre-feet of water per year from the Bureau of Reclamation's Colorado–Big Thompson Project (C-BT) to irrigators, rural water districts, large industries, and towns in northeastern Colorado (see Tyler 1992 for a detailed history). The District charges irrigators about $3.50 per acre-foot for water delivered to their shares in the District (so-called allotments). Cities and industry pay about $7.00 per acre-foot. The remaining costs of the District, including the repayment of the capital costs of the original C-BT Project, are paid from real estate taxes levied against all urban and rural real estate in the District.

The cities that use C-BT water have various pricing schemes. Fort Collins residential users are mostly unmetered and pay a flat fee of $17.47 plus $0.12 per 100 square feet of lot area per month. Boulder has an increasing block rate structure for residential users of $1.16 per thousand gallons for the first 6000 gallons, $1.42 per thousand for the next 15,000 gallons, and $2.56 per thousand after that. In addition, there is a waste-water treatment fee that depends on winter water use. Longmont is partly metered and partly unmetered.

There is an active market for shares (allotments) in the Northern District. The market is limited to the boundaries of the District, with a current market price of about $1800 per share. Since the shares, on average over the years, deliver 0.7 acre-feet per share per year, this implies a price of about $2600 per acre-foot for a perpetual supply or perhaps about $208 per acre-foot per year using an 8 percent interest rate. In addition, there is an annual market for the temporary use of C-BT water—the "rental" market in which shareholders sell excess water to other users on a one-year basis. These "rental" prices range from $7.50 to $25.00 per acre-foot, depending on snowpack and rainfall conditions and time within the growing season, usually being higher later in the season.

So, what is the price of water in northeastern Colorado? There are many different prices, depending on what type of water user you are, where you are located, and so forth. The real estate tax constitutes a large part of total revenue for the District but is surely not a price for water. The correct definition of price should be "the amount paid for the *next* (or marginal) unit withdrawn." This is the correct definition because it is the *behaviorally relevant* measure of the cost incurred by the water user in using one more unit. It is that cost that a rational user will compare with marginal benefits in deciding how much water to apply.

## Types of Decisions Affected by Water Price

In the short term when water users' technologies and markets are fixed, the price of water (as defined earlier) will affect the quantities withdrawn from the supply system. If your lawn watering is about to move you into Boulder's most expensive block, and *if* you are aware of that, it may induce you to control your water applications more carefully. This decision will not be affected by other dimensions of the total "financial package" (or revenue structure) such as monthly fixed charges, charges per 100 square feet of lot area, sewage fees, or up-front hookup fees.

In the longer term, other types of decisions *will* be affected by these other dimensions of the financial package. Under the Fort Collins charge of $0.12 per 100 square feet of lot area per month, some house buyers might seek smaller lots, or if the lot area refers to *irrigated* area, many households may turn to xeriscaping or other ways of reducing the irrigable area. In Denver, the current hookup charge for a typical single-family residence is about $5000, whereas the fee for an apartment is $2800. Over time, this cost difference will affect decisions on types of housing to be built and will thereby affect the structure of the city.

The farmer applying irrigation water to corn or alfalfa within the Northern District might mistakenly consider the price of water to be the District's annual charge of about $3.50 per acre-foot, but all farmers today are astute business managers and would immediately understand that they are foregoing short-term (rental) revenues of up to $25 per acre-foot and the possibilities of longer-term revenues (from permanent water sales) of as much as $208 per acre-foot.

Thus, the relevant measure of *cost* to the water user depends on the question being asked. In some cases, it is the price of water (as defined earlier) and in other cases the relevant cost contains other elements of the water supply "financial package."

## The Conflicting Roles of Water Price

From the viewpoint of the efficient allocation of resources over time, the primary function of water price is the efficient allocation of existing supplies in the short run and the provision of needed information for the optimal expansion of supply capacity over time. In the short run, the delivery capacity of a supply system is fixed. Since price affects the quantities users demand, price can be used to adjust demand to the available supply. During drought periods when demands are high and supplies low, price can be raised to reduce demands. Variations on this theme would include seasonal pricing with higher summer prices, reflecting the higher costs of providing the higher flow rates demanded and time-of-day pricing to encourage users to diversify the timing of their demands during the day.

Why is price a good way of rationing water among users under fixed delivery capacities? We can assume that water users compare marginal benefits with marginal costs when deciding on changes in water use patterns. If all water users (of a particular class of user, say, residential users) face the same price, each will adjust water use until the marginal benefits fall to equality with the price. All users in that class then exhibit the same marginal benefits—a commonsense necessary condition for maximizing the total benefits from the use of available water. Naturally, there are other ways of rationing the available supply, such as requesting cutbacks, requiring even-odd day watering, or prohibiting certain uses. While these rationing mechanisms often work, none of them guarantees that water will be distributed to the users who place the greatest value on the water, nor can users choose what types of use are more valuable to them. Naturally, some people will object on equity grounds to raising price during periods of shortage, but devices such as "lifeline" rate structures with a low price per thousand gallons for the first few thousand gallons can remedy these situations.

In the longer term, price *may* be of use in helping to determine optimum supply system expansion over time. Looking again at the Northern District in Colorado, if water demands in the region grow, the price of District shares will rise. That price indicates the marginal value of added supply to water users. If the price rises to equality with the unit cost of the best available supply project, it could tell us that the new project is now warranted.

In fact, there are not too many situations where the price of water is free to respond quickly to changes in the demand-and-supply situation. In urban systems, the price structure is determined by management and only infrequently adjusted to changing demand and cost conditions over time. In some rural areas, the market for irrigation water is greatly restricted or even nonexistent because the water supplies are from state or federal projects that contract with irrigation districts and price the water to cover some parts of historical construction costs and current O, M & R costs.

Another major function of price is the production of revenues for the water supply agency. Naturally, all components of the "revenue package" contribute to revenues. Whether increases in price (as defined earlier) will produce more or less revenue depends upon the responsiveness of users to price. Economists call this responsiveness the *price elasticity* of the user's demand for water. There is a vast literature on the price elasticity of water demands in residential, industrial, and agricultural uses (see Foster and Beattie 1979; Billings and Agthe 1980; Howe 1982; Boland et al. 1984; Schneider and Whitlatch 1989; Nieswiadomy 1990; Tate 1990; Renzetti 1992). The consensus is that residential and industrial demands (except for cooling water) are "inelastic" while agricultural demands are "elastic." *Elasticity* is defined as the percentage change in quantity demanded divided by the percentage change in price. It follows that, for "inelastic" demands, if price is raised, revenue collections also will rise while for "elastic" demands, an increase in price will reduce revenues. Thus, knowledge of demand elasticities is a key to financial planning for water utilities.

Another consideration in water pricing is fairness or equity among water users. It is often asserted that raising the price of water or charging for it where it has historically been free will "hurt the poor" or be unfair to them. This may be a substantive issue in Third World countries where much of the population is at subsistence level, but in advanced countries, the fairness issue can be handled through "lifeline pricing" (as defined earlier). This is not to say that poverty isn't a problem in advanced countries, but lowering the price of water to all users is an inefficient way of dealing with the problem.

## Pricing and Water Quality

The price of irrigation water greatly affects the quantity of water applied to crops. It thus affects the volumes of runoff and deep percolation. Underpricing of water can thus lead to water logging of the soil, drainage onto lands at lower altitude, and the leaking of toxics and heavy metals from the soil and subsoil. This is precisely the story of the Kesterson Reservoir in California which was poisoned by selenium from the drainage waters of the Westlands Irrigation Project (see National Research Council 1989). The State of California and the Bureau of Reclamation wanted to treat this as a drainage problem while the real origins of the problem were too cheap water (far below cost) from the (federal) Central Valley Project and the large agricultural subsidies that made it privately profitable to open up irrigated lands that weren't appropriate for irrigation and weren't needed in the face of surplus crop supplies nationally.

In the administration of water quality, once again water pricing is important. A town uses water and returns a fraction of it to the municipal wastewater treatment plant. The treatment plant, after primary, secondary, and perhaps tertiary treatment, returns the treated water to the river. How is pricing relevant to this cycle of water use and waste disposal?

First, prices charged for residential, commercial, and industrial water supply will affect the quantity of water withdrawn and thereby the volume that returns through the sanitary sewer for treatment. Charges are also made to cover the costs of wastewater treatment, but these are usually based on average winter use (nearly all of which returns to the sanitary sewer) so are unlikely to affect current use. Nonetheless, the volume of wastewater flow will affect the concentration of pollutants and thus treatment costs. While residential users have little freedom in adjusting their waterborne waste loads, commercial and industrial water users have a wide range of technologies, input and product choices that affect their waterborne waste loads. Thus, charges on the waste flows from commercial and industrial establishments (i.e., prices for pollutants) will induce changes to lighten the pollution load. Toxics and heavy metals probably should be prohibited rather than priced.

In the final step of this chain of events, the treated effluent is poured into the river and the sanitized sludge is applied to farmland. The nutrients that enter the river cause damages to other water users, and these damages should be reflected in charges (taxes) against the outflow of nutrients.

Such taxes have been widely used in Europe but not in the United States. The farm application of sludge benefits the crops and in many situations is the least-cost method of disposal. If farm demands for sludge exceed supplies, the utility could establish a price for it that would equate supply and demand and produce revenue for the utility.

## Pricing and Water Supply Reliability

The reliability of urban water supply is an important attribute of a supply system. Moderate, short-lived shortages can be accommodated by the water-using public with only minor inconvenience. Longer-term failure to supply water as demanded during the summer can lead to loss of valuable perennials, bushes, and trees. (The estimated loss per household in Marin County, California, during the 1976–77 drought was several hundred dollars. See Nelson 1979.) Extreme shortages requiring reductions of in-house and public uses can endanger health.

Reliability of urban supplies depends on two major system components: raw water supplies and water infrastructure (treatment, distribution, etc.). Raw water reliability can be increased through the acquisition of more water rights, more storage, melding sources from different watersheds and groundwater, and so forth. Infrastructure reliability is enhanced by good maintenance, equipment replacement, and redundancy. Each of these steps is costly. The problem is to value changes in reliability so that incremental benefits and costs can be compared. But there is no market for reliability and thus no explicit price for it. Its value is a nonmarket value or price that must be determined by indirect methods.

Surveys of water customers' willingness to pay increased monthly bills for increases in reliability (or willingness to accept a lower bill in return for reduced reliability) have been used to value reliability changes. In a study of three Colorado towns (Howe and Smith 1993; Howe, Smith, et al. 1994), reliability was defined as the complement of the annual probability that a shortage requiring curtailment of outdoor use would occur (i.e., 1.0 minus the probability of such a shortage). Boulder exhibited a shortage probability of only 1/300, while rapidly growing Aurora and Longmont had shortage probabilities of 1/10 and 1/7, respectively.* The majority of Boulder respondents preferred a *decrease* in reliability (specifically an increase in

---

* In spite of Boulder's high estimated reliability, the system nearly broke down one day in February, 1992 during the study. Two raw water pipelines clogged with ice within the same hour and attempts to draw water from Boulder Reservoir were stymied by a gate valve that broke.

the probability of shortage to 1/100) in exchange for a reduction of $4.50 in their monthly water bill. Surprisingly, near majorities in the other two towns *also* preferred decreases in reliability in return for monthly bill reductions.

This is an example of estimating an important nonmarket value for planning purposes, that is, placing an estimated "price" on a finite change in water-supply reliability. Similar techniques have become important in placing prices on air and water quality changes, additions to wilderness areas, and other nonmarket amenities.

## Conclusions

We see that water prices, appropriately set and applied at different points of the water supply and use cycle, perform many valuable functions, namely to allocate existing supplies efficiently by confronting water users with the costs of providing water, to help signal water suppliers when supply augmentation is needed, and to help shape a rational approach to a healthy water environment.

Many pricing improvements need to be made. Irrigation water and wastewater disposal need to be priced more in keeping with the total costs involved. City water users should be metered and, in most cases, charged according to an increasing block rate schedule to reflect the increasing scarcity of raw water. Urban hookup fees should be raised to a level reflecting current raw water development plus treatment and distribution investment costs.

The regulations surrounding water pricing by supply agencies need to be reconsidered to allow more appropriate levels of prices and greater freedom to change price under changing conditions. Most utilities have a zero-profit constraint. During drought, this means that raising prices may be ruled out. It may also preclude building up sinking funds for future expansions. Many cities in the western United States acquired their raw water supplies (water rights) long ago and their value is not accounted for. The zero-profit constraint then means that the value of the raw water being used is not reflected in the water prices charged. City accounting practices should be changed to carry those water rights on the books as *assets* and to count as a current cost reasonable interest charges on the value of those rights.

A final needed change relates to undoing or overcoming the effects of past inappropriate pricing of water. This is especially needed in the case of water and electric power provided by federal projects (Bureau of Reclamation and Corps of Engineers) where water is distributed to irrigation districts under long-term, highly subsidized fixed-price contracts. The resultant underpricing of water is reflected in higher farm profits and hence into higher farmland prices. The same is true for the users of underpriced federal power. Any change in pricing policy, including the changes that *should* occur from an efficiency point of view, will be translated into changes in farm and business profits and land values. Naturally, such changes will be resisted by the beneficiaries and their political representatives.

It has been argued that this historical underpricing of water was an intentional policy of the national government that should not be changed to the detriment of the intended beneficiaries without some compensation. However there is no need to hold up the rational reallocation of water (say from agriculture, where it is worth $30 per acre-foot at the margin to growing urban and recreational uses where marginal values are reflected in willingness to pay for new sources of water of $400 to $800 per acre-foot) because of equity concerns. *Water markets* will allow current water contract and water right holders to decide whether to sell some or all of their water. Many will decide to sell and will thereby receive the *capitalized value* of the subsidies created by their contracts. The water will then move to better uses and water owners will receive full compensation (see Wahl 1989). There are hundreds of such win-win opportunities in water pricing in the United States today.

## References

Billings, Bruce and Donald E. Agthe. 1980. "Price elasticities for water: a case of increasing block rates," *Land Economics*, February.

Boland, John J., B. Dziegielewski, D. Baumann, and Eva Optiz. 1984. "Influence of Price and Rate Structures on Municipal and Industrial Water Use," a report submitted to the U.S. Army Corps of Engineers, Institute of Water Resources, June.

Foster, Henry S. Jr., and Bruce R. Beattie. 1979. "Urban Residential Demands for Water in the United States, *Land Economics*, February.

Howe, Charles W. 1982. "The Impact of Price on Residential Water Demand: Some New Insights," *Water Resources Research*, 18(4), 713–716.

Howe, Charles W. and Mark Griffin Smith. 1993. "Incorporating Public Preferences in Planning Urban Water Supply Reliability," *Water Resources Research*, vol. 29, no. 10, pp. 3363–3369.

Howe, Charles W., Mark Griffin Smith, et al. 1994. "The Value of Water Supply Reliability in Urban Water Systems," *Journal of Environmental Economics and Management*, 26, pp. 19–30.

National Research Council, Committee on Irrigation-Induced Water Quality Problems. 1989. *Irrigation-Induced Water Quality Problems*, Washington, D.C.: National Academy Press.

Nelson, John O. 1979. "Northern California Rationing Lessons." In American Society of Civil Engineers *Proceedings of the Conference on Water Conservation*, New York: ASCE, pp. 139–146.

Nieswiadomy, Michael L. 1990. "Estimating Urban Residential Water Demand: The Effects of Price Structure, Conservation and Education," paper presented at the Western Economic Association Meetings, San Diego, June.

Renzetti, Steven. 1992. "Estimating the Structure of Industrial Water Demands: The Case of Canadian Manufacturing," *Land Economics*, November.

Schneider, Michael L. and E. Earl Whitlatch. 1990. "User-Specific Water Demand Elasticities," *Journal of Water Resources Planning and Management*, July.

Tate, D. M. 1990. "Water Demand Management in Canada: A State-of-the-Art Review," Social Science Series No. 23, Inland Waters Directorate, Environment Canada, Ottawa.

Tyler, Daniel. 1992. *The Last Water Hole in the West: The Colorado–Big Thompson Project and Northern Colorado Water Conservancy District*, Niwot, CO: University Press of Colorado.

Wahl, Richard W. 1989. *Markets for Federal Water: Subsidies, Property Rights and the Bureau of Reclamation*, Washington, D.C.: Resources for the Future, Inc.

# 7

# Phoenix Changes Water Rates from Increasing Blocks to Uniform Price

## William R. Mee Jr.
### City of Phoenix Planning Department

This case study is intended to provide an example of how technical analysis and economic theory are used as a basis for political decisions about how the cost of water service should be recovered from customers. Conversely, it is an example of how the development of water tariffs is as much or more influenced by public decision-making processes and corporate culture as by theory and analysis. Some public interests have more influence than others, and some staff and decision makers are more willing to take risks than others. What will be described is a process of starts and stops over several years leading to a radical change in the Phoenix water rate structure, an analysis of that structure, and a failed attempt to further improve it. The current Phoenix water rate structure may appear to be more evolutionary than radical, but the changes made were not only contrary to standard industry practice but were also a shift away from then recent trends becoming more common in the industry.

The case study begins with a brief description of Phoenix and why it embarked on water rate structure revisions. It then provides a description of prior rate structures. The next part of the case study presents the process of studying alternative rates and a description of the current rate structure. It concludes with concerns about the current structure and proposals for improving it.

## Information About the Phoenix Water System

Phoenix is located in the Salt River Valley in the northern reaches of the Sonoran Desert. The City provides water service to over one million people and has a water service area of 600 square miles. Summer temperatures often exceed 100 degrees Fahrenheit and average rainfall is seven inches per year.

Growth of Phoenix and surrounding cities was first supported with wells and water from the country's first reclamation project, the Salt River Project, and beginning in 1986 with water from the Central Arizona Project. These sources coupled with modest groundwater use now provide a secure but increasingly expensive water supply. Improved public health requirements and expansion to meet the needs of a rapidly growing city also have caused significant increases in the costs to supply water.

Because of the hot, arid climate, water demand increases substantially in the summer. Single-family homes consume about half of all water delivered and use three times as much water in the summer as in the winter. A third of water is used by nonresidential customers and the remainder by multifamily customers. (See Figures 7-1 and 7-2.)

## Genesis of Interest in Changing Water Rate Structures

In 1986, the City of Phoenix adopted the *Water Conservation Plan 1986* that recommended implementation of 15 demand management measures. The highest implementation priority of these measures was a study to develop and evaluate a water rate structure based on marginal cost pricing. The 15 measures and their priority were based on a study begun in 1985 that focused on quantifying the benefits of various water demand management measures. Three factors were used to evaluate potential measures: economic feasibility, technical feasibility, and social acceptability. After a selection of measures likely to be appropriate for Phoenix, cost-benefit analyses

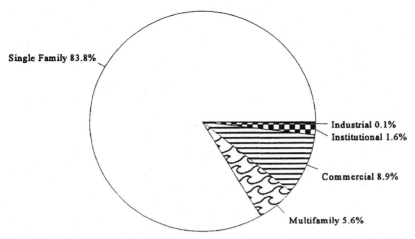

**Figure 7-1.** Water Accounts by Major User Groups, 1988–89

were performed to determine cost effectiveness. A literature search and review of current laws and ordinances were conducted to evaluate technical feasibility. In addition, a survey of Phoenix citizens, along with detailed interviews with community leaders and prominent business and agency executives, provided an indication of each measure's social acceptability.

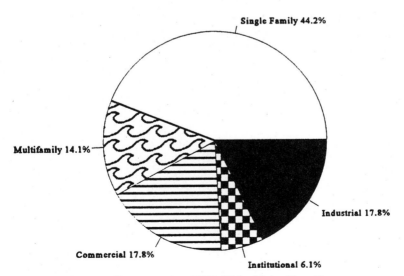

**Figure 7-2.** Water Consumption 1988–89 by Major User Groups

Results of the analysis indicated that most of the conservation measures studied would yield benefits both to the City government and to the citizens of Phoenix. Benefits included water savings, reduced groundwater pumping costs, reduced or delayed capital improvement costs, and reduced costs through energy savings.

## The Previous Water Rate Structure

Phoenix has used several different rate structures since it started operating a water utility in 1907. By the 1970s the rate structure had become a mildly increasing block form with uniform charges throughout the year. Following a comprehensive water rate study in 1981, the structure became more complex and more steeply increasing. It consisted of three customer classes—residential, commercial, and industrial—and was separated into a summer season (May through October) and a winter season (November through April). For each customer class, the rate structure also included three rate blocks with different rates for 0 to 10 hundred cubic feet (CCF), 11 to 25 CCF, and greater than 25 CCF. The per CCF consumption charge increased with each higher block. In addition to the consumption charge, a monthly service charge based on meter size was assessed on each account. The consumption and service charges are summarized in Table 7-1 for accounts inside the City of Phoenix.

The increasing per unit cost in each block of the previous rate structure did provide an incentive for customers to maintain their demands below 25 CCF per month. However, residential customers in the lower blocks and the majority of the nonresidential customers were actually being provided water at below the then annual average unit cost of $0.77 per CCF. This below-cost pricing was assumed to be encouraging these low-use customers to use water inefficiently.

Water conservation was encouraged by the seasonal design of the previous rate structure. The residential sector faced higher costs of an additional $0.32 per CCF for the top block in the summer. Industrial customers pay a $0.79 surcharge for every CCF above their winter average. The higher summer charge was designed to impact the more price-responsive outdoor water demands and to decrease summer peaking. This charge, however, was assessed against a very small proportion of total water use.

Although the increasing block rate structure was intended to encourage water conservation for high water-using customers, few customers were likely to understand the rate structure. The rate structure was also difficult for Water Customer Service staff to explain to customers.

**Table 7-1.** Previous Rate Structure

| Consumption block in 100s of cubic feet | Monthly price per 100 cubic feet by customer class | | | | | |
|---|---|---|---|---|---|---|
| | Residential | | Commercial/ government | | Industrial | |
| | Winter | Summer | Winter | Summer | Winter | Summer* |
| 0 to 10 | $0.42 | $0.42 | $0.42 | $0.42 | $0.49 | $0.49 |
| 11 to 25 | $0.58 | $0.68 | $0.55 | $0.57 | $0.62 | $0.62 |
| Over 25 | $0.92 | $1.24 | $0.73 | $0.85 | $0.72 | $0.72 |

| Monthly service charge | |
|---|---|
| Meter size (in inches) | Charge |
| ⅝ by ¾ | $4.70 |
| 1 | $11.94 |
| 1½ | $16.42 |
| 2 | $28.06 |
| 3 | $51.87 |
| 4 | $68.07 |
| 6 | $94.04 |

* A surcharge of $0.79 was assessed to industrial customers for every unit of summer consumption above their winter time average.

The seasonal aspect of the previous rate structure was justified on a cost basis, but was also intended to reduce summer water demand. Unfortunately a customer survey prior to the adoption of the new water rates found that only 50 percent of the Phoenix water customers were aware that the summer rates were higher than winter rates.

## Study of Alternative Water Rate Structures

Following adoption of the Water Conservation Plan, staff began working to implement each of the recommendations including the one for a water rate structure study. At the same time the Water Services Department financial staff had identified a need to increase water rate revenues and to update models used to calculate water rates. Requests were sent to consultants with water rate expertise, asking for proposals on how they would address the three objectives of revenue requirements, rate structure improvements, and rate model development. It was found that few water rate consultants had experience in developing alternative water rate designs and none had experience with use of marginal cost pricing for water utilities. Even though there was growing use of marginal cost pricing by electric utilities, little of what had been learned there was transferable to water utilities. The consultant selected had extensive experience in calculating water rates for public utilities and experience in rate structures other than the American Water Works Association standard.

Staff and consultants studied a wide variety of water rate structure options during a parallel effort to develop a new water rate calculation model. The model was needed both to improve normal calculation procedures and to develop a capacity to translate revenue requirements into alternative water rate structure options designed to collect the same revenue.

The following objectives were established to guide evaluation of proposals for alternative water rate structures:

For any rate structure to be successful, it must:

- Raise sufficient revenue to pay the costs (REVENUE SUFFICIENCY)

- Apportion charges according to responsibility for costs (EQUITY)

- Encourage the economic optimum use of the service by the customers (EFFICIENCY)

- Be simple enough to be understood and accepted by the public and not require constant revision (SOCIAL ACCEPTABILITY and PRACTICAL FEASIBILITY)

- Encourage WATER CONSERVATION.

To encourage water conservation it was determined that a water rate structure and the rates should meet the following criteria:

- Prices should be adjusted to at least rise with inflation.

- Consumption charges should be emphasized over fixed charges.

- Small users as well as large users should be charged the same unit price for water.

- Unit prices should be based on marginal costs.

The previous rate structure did provide sufficient revenue. However, revenues were difficult to predict since prior knowledge of where every customer will be in the rate structure for a given billing period is required. Equity is also difficult to determine since every customer must be examined individually by month to make this determination. Finally, the lack of simplicity did not allow the potential for water conservation to be fully realized for many customers.

Several staff and consultant brainstorming workshops were held to develop proposals for meeting the objectives and the following alternatives were carried forward for additional analysis:

- Current water rate structure
- Current with an additional block for high residential use and the expansion of the industrial surcharge to other nonresidential uses
- Uniform price set to recover marginal costs
- Uniform price set to recover historical costs
- Peaking, where summer rate blocks were set on the basis of summer-to-winter water use ratios
- Modified block (described below)

Options were also discussed for adding different rates to different parts of the Phoenix water service area to some of the structure alternatives to reflect differences in the costs of water and treatment. After study of some of the more significant changes previously listed, it was decided to take an incremental approach with the modified block option. Staff and consultants proposed a rate structure alternative with the following characteristics:

- Change the summer period from 6 to 5 months.
- Change the fixed base charge to a minimum charge (substantially reducing fixed revenues).
- Change the size of rate blocks to correspond to use patterns by customer class.
- Charge all customers at a unit price that reflects the average historical cost of water.
- Charge customers with above-average use a higher unit price reflecting some additional future costs.

Even this incremental change proved to be too significant to consider at a time when the city council was being asked to increase revenues, so in early 1988 the two issues were separated. Council gave conceptual approval to the rate structure proposal, but deferred final action.

Because of the complexity of both the water rate revenue requirements and the water rate structure issues, city staff recommended that the council appoint a citizen's committee to advise them on water rates. This approach had been successful in Tucson, Arizona, and an ad hoc committee was quickly appointed by the Phoenix city council. The committee was asked to review staff's request for additional revenue. After intensive work by staff, consultants and the committee, the council adopted increased water rates on April 1, 1988.

Staff then recommended that a permanent Citizens' Water Rate Advisory Committee (CWRAC) be appointed. The council agreed and appointed the committee which held its first meeting on August 24, 1988. The committee began working on water rate structure alternatives as well as a fiscal year 1989–1990 revenue requirement. During the fall and winter of 1988, staff and consultants provided training for committee members on both the technical and theoretical basis for water rate setting and rate structure design. This included two day-long workshops.

The original CWRAC spent nearly three years from 1987 to 1990 evaluating several different options based on the rate objectives. After the long review, the CWRAC decided that a rate structure that achieved the following goals was desired:

- Increased the conservation incentive over previous rate structures
- Simplified the structure for customer understanding and for forecasting annual revenues
- Maintained stable bills for low water use customers
- Moved away from embedded or historically based rates and toward marginal cost pricing
- Eliminated subsidies between customer classes unless data supports cost of service differences
- Reflected any seasonal cost of service difference in the rates

The committee then began to focus on the water rate structure alternatives. Staff reviewed the full range of options previously considered. Considerable time was spent on the peaking and modified block options. Under the scrutiny of the committee the deficiencies of these options became obvious. In addition to administrative difficulties, the peaking option would have resulted in substantially different charges for similar use during the summer based on differences in historical use. At first it seemed fair to charge more to people with large differences between winter and summer use. The city's economic consultant, however, was able to convince staff and the committee that peak demand was caused by anyone using water during a peak period and that all peak-period use should be charged the same.

The staff and consultant recommended modified block option also failed the committee's review. Staff had done such a good job convincing them that there was no equitable way to set rate blocks because of the variety of water use situations, the committee rejected all use of blocks.

Customer classes were also rejected because staff and consultants were unable to show differences in costs to serve classes. Normal rate setting practice used hourly and daily peaking to differentiate among customer classes. Even if class averages could be used, the data on hourly and daily peaking by customer class were limited and felt not to be sufficiently reliable. More important, customer water meters do not record hourly or daily use, and therefore individual customer behavior cannot be influenced by rates based on hourly and daily peaking.

This work led to the development of the current rate structure. The first proposal reviewed had the following characteristics:

- No customer classes
- No rate blocks
- Four seasonal rate periods
- A fixed monthly charge based on the cost of reading and maintaining the meter

The first rate developed showed a unit charge for water significantly below the charge in the top block in the then-current structure. Since the top block charge was already less than estimates of summer period marginal costs, it was decided that the new rate structure summer charge should be at least similar to the current charge.

The technique used to raise the unit cost of water was similar to the minimum monthly charge concept proposed in the modified block alternative. On an annual average, the monthly charge for a ⅝-inch meter (representing 95 percent of all meters) was equal to about the unit charge for 6 CCF of water. It was thus decided to include 6 CCF of water in the fixed monthly charge at no additional cost. Staff assumed, as later analysis showed, that there was limited opportunity to cut water use with a variable charge for the first 6 CCF. A higher unit charge for use above 6 CCF was assumed to cause a valuable net reduction in water use even with some expected increase in the water use of customers using 6 CCF or less. This change resulted in a higher unit charge, but summer rates were still lower than the then-current (now called previous) rate structure. To further increase the summer unit price, the amount of water included in the base charge at no additional cost was increased to 10 CCF. During the summer in Phoenix, outdoor water use of the additional 4 CCF or less was also assumed to be relatively fixed and not subject to much increase even if provided at no cost.

As the rate structure was further evaluated, the no-charge block began to be called a *lifeline*. The pattern of monthly water use in a hypothetical water-conserving home (ultra-low-flow plumbing fixtures and low-water-use landscape) resulted in water bill decreases except in the spring season.

This originally unintended benefit became a significant selling point for the new structure.

The current rate structure has the following characteristics:

- No blocks or customer classes

- Three seasonal periods

- A lifeline of 6 CCF from October through May and 10 CCF from June through September

- Varied base charge by meter size to cover meter reading and maintenance

## The Current Water Rate Structure

Table 7-2 illustrates the rate structure and prices when it was adopted on June 1, 1990.

Compared to the previous structure, the new structure is easier to understand and revenue forecasting is simpler. It also meets the conservation objective, estimated to cause a 3.8 million gallons per day reduction in water use. The higher price for summer water use was projected to flatten peak day demand by 3.6 percent. A modest increase in indoor demand was projected as a result of no charge being levied for the lifeline amounts of water.

**Table 7-2.** Current Rate Structure (as of June 1, 1990)

| Consumption block in 100s of cubic feet | Monthly price per 100 cubic feet by season for all customers | | |
|---|---|---|---|
| | Winter | Spring/fall | Summer |
| 0 to 6 | $0.00 | $0.00 | $0.00 |
| 7 to 10 | $0.74 | $0.90 | $0.00 |
| Over 10 | $0.74 | $0.90 | $1.17 |

| Monthly service charge | |
|---|---|
| Meter size (in inches) | Charge |
| ⅝ by ¾ | $5.12 |
| 1 | $6.06 |
| 1½ | $7.47 |
| 2 | $8.41 |
| 3 | $25.99 |
| 4 | $25.99 |
| 6 | $33.25 |

## Impacts of Current Rate Structure on Customer Classes and Selected Customers

With any rate structure change, some customer bills will increase and some decrease. Because the monthly water use pattern of an individual customer greatly affects the annual bill increase or decrease a customer may experience, annual water bill changes for customer classes are difficult to determine. For example, Figure 7-3 shows the proportion of single-family customers receiving a higher or lower monthly water bill as a result of the rate structure change. As can be seen, the majority of customers, but not all, saw bill decreases in the winter months. Conversely, the majority, but again not all, saw bill increases during the remainder of the year. This reflects Phoenix residential water use patterns where the majority of customers have significant outdoor water use. Some customers, however, have little outdoor water use and therefore had lower summer bills due to the larger summer lifeline amount.

Large commercial and industrial customers using over 1000 CCF per month experienced the largest increases in their summer bills with up to a 37 percent increase in summer water bills.

## Summary of Current Rate Structure Benefits

The current rate structure removed many of the inequities of the previous rate structure. The increased simplicity, in conjunction with the proper

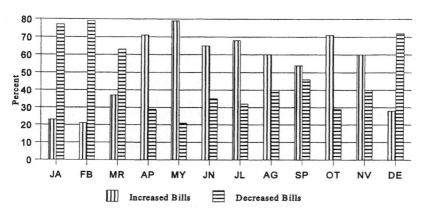

**Figure 7-3.** Impact of Bill Structure Change on Single-family Accounts

price signals, did allow consumers to respond with more efficient use of their water resources. The estimated water use reductions of 3.8 million gallons per day allows Phoenix to avoid the cost of chemicals, electricity, and operation for water and wastewater treatment. Over a fifty year period, this is over $7,000,000 in present value benefits for the avoided treatment costs. This reduction also allows deferral of construction of water and wastewater treatment plants and even larger avoided costs.

The new water rate structure improved the equity of water pricing and provided an increase in the proportion of water customers receiving a price signal that encouraged water conservation.

## An Attempt to Change the Water Rate Structure Again

The current water rate structure has been well received by the water industry as an effective water conservation rate structure. However, in hindsight there are some changes that could make the rate structure even more effective in meeting rate objectives.

In addition to Water Services Department staff interest in changing the water rate structure, the Arizona Multihousing Association lobbied staff, the committee, and the city council to change the structure. The association represents the interests of owners and managers of large apartment projects. These large projects, like other large water customers were adversely impacted by the 1990 water rate structure change. Their summer unit rate for water increased. More significantly, though, they felt they were being unfairly treated because the lifeline is applied to water accounts rather than housing units. Each single-family home gets a lifeline, but water meters at apartment complexes typically serve 10 to 20 apartments. This results in the average unit price for water to apartments being higher than to single-family homes, even though all water over the lifeline has the same price.

The Association's lobbying resulted in direction to staff to study ways to address the concerns raised. Since there was no interest in bringing back customer classes, there was no easy way to "fix" the apartment problem without changing the way all customers were charged. Staff developed several rate structure options focused on reducing the influence of the lifeline on the average price of water. In addition, staff was working on changes to the way water rates were calculated to reflect costs of service more accurately and to move toward seasonal marginal costs.

Discussion of the rate structure options with the committee started with proposals made by representatives of the Association and a representative of coin-operated laundries to eliminate the lifeline. As will be seen in the following, even reducing the lifeline by half causes significant percentage increases in summer water bills for the majority of single-family customers. Since, as was shown in Figure 7-1, single-family water accounts are 84 percent of all water accounts, eliminating the lifeline was viewed as politically impossible. The committee decided to reduce the lifeline and call the reduction the first phase of a rapid elimination of it. The following describes in more detail concerns about the 1990 rate structure, the committee's recommended rate structure alternative and a cost of service update provided by staff. It also provides an analysis of how the proposals meet the water rate objectives.

## Concerns with the Current Rate Structure

Before addressing Water Service Department recommendations for the additional water rate structure changes, a brief discussion of complaints by customers may illustrate some refinements that should be considered.

### Customers Below Lifeline Cannot Impact Bill by Conserving

The block of consumption included in the service charge or "lifeline" provides the lowest possible water bills to customers below the lifeline allocation. These customers are just paying for the cost to read and maintain the meter and to generate a bill. Other costs of water service are reflected in the volume charge assessed to consumption above the lifeline allotment. Any customer that maintains water use within the lifeline allocation will experience the same bill every month. Some customers have complained that they have made efforts to reduce their water use but noticed no impact on their water bill. A policy decision made in 1990 was to develop a rate structure that provides stable bills to low water use customers. The objective of stable or constant bills for small customers removes the incentive to conserve for economic reasons.

### Customers with Nonseasonal Demand Experience Lower Summer Bills

A goal of the current water rate structure is to reflect the changing cost of providing water to the customer throughout the year in the customer's monthly water bill. The three season rates, with the summer months higher than the winter months, communicates this message. However, the increase in the lifeline allocation from 6 CCF in the medium season to 10 CCF in the high season overwhelms the change in rates. The change from 6 to 10 CCF represents a 40 percent change while the rate increases from $0.95 to $1.24 is a 30 percent change. Unless water use increases by more than 10 percent between the medium- and high-rate seasons, the water bill will actually decrease in the high season.

Figure 7-4 illustrates a small single-family customer with very little landscape or seasonal water use whose water bill actually decreases in the summer months. This has been a common complaint by customers who receive a bill notice in May telling them to expect their summer bill to go up and then they actually experience a bill decrease. The best method to deal with this complaint is to make the lifeline allocation equal in all months of the year.

## The Effectiveness of the Current Lifeline in Assisting Low-Income Customers

The term lifeline as used in Phoenix water rates is somewhat of a misnomer because of the interpretation that it provides a direct benefit to low-income customers. The lifeline really provides a benefit to low water use customers, not necessarily low-income customers. Surveys and water use studies suggest that on average higher-income households use more water than lower-income households. This is because low-income residents are less likely to maintain high-quality landscapes and to own water-intensive appliances such as a swimming pool or spa. However,

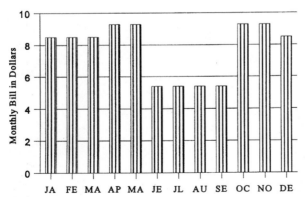

**Figure 7-4.** Water Bill Pattern for Low, Uniform Single-Family Use

this is only on average. Some low-income customers also have leaking pipes or poorly maintained evaporative coolers that make their water use very high. Evaporative coolers, which are a major user of water in the dry summer months, are less likely to be maintained properly on older homes where low-income residents reside.

As an extreme example to illustrate this point, two adjacent households are compared. In one household is a single mother with a low-paying job, five children, a grass landscape, leaking plumbing, and a poorly maintained evaporative cooler. In the house next door is a single professional with a high income, a well-maintained home, and a low water use landscape. The single professional maintains his water use below the lifeline in all months while the low income household greatly exceeds the lifeline amount. As mentioned previously, the so-called lifeline aspect of the water rate structure benefits low water use customers, not necessarily low-income customers.

True lifeline rates are rates that are set below costs for all levels of use and provided to qualifying low-income customers. The City of Los Angeles administers a low-income and elderly rate that is below the rate paid by other customers. The block-of-consumption type of lifeline in Phoenix has been termed a "shotgun" approach to assist low-income households, because it benefits only low-income customers with low water consumption.

### The Negative Impact of the Lifeline on Large Water Use Customers

A complaint by large water use customers is that by providing the lifeline block of consumption in the service charge, they pay higher water bills. It is true that the lifeline block of water consumption does not generate any revenue and therefore pushes the rate up on the remaining consumption. A rate is simply total costs, less revenue generated from the service charge, divided by water consumption. As was originally intended, not billing all of the consumption causes the variable rate to go up to recover the revenue. The resulting rate can be viewed as fair, since all customers receive the lifeline allocation. However, as the lifeline allocation becomes a smaller and smaller portion of the customer's total water demand, the volume charge becomes more significant in determining the total water bill.

Modifying the current rate structure to contain no lifeline allocation would result in a lower volume charge with the larger users paying less. Conversely, small users currently within the lifeline allocation would pay substantially more (in proportion to their current bill). In some cases bills for small water users would more than double. The lifeline allocation, as

mentioned previously, is provided to all customers but specifically benefits customers whose consumption is not substantially above the lifeline.
It is a part of the rate that by policy provides stable bills to low water use
customers.

## Water Services Department Concerns and Recommendations

### Cost-of-Service Update

A cost-of-service update is simply a study to determine what costs need to
be recovered in the water rates and to discern if those costs are recovered
in a manner that reflects the cost of service policy of the water utility. The
American Water Works Association (AWWA), the water industry body
which sets standards for most functions related to water utilities, recommends that the utility consider its operations, economic objectives, and its
customers when selecting a rate structure. For the Phoenix water service
area, the higher summer demand and the high cost for plant capacity
expansion and new supply acquisition have resulted in the current rate
development policy. The policy or rate philosophy is to allocate all of the
capacity costs not used in the winter to the summer or high season when
the extra capacity is used. This results in the difference between the high
and low season rates and communicates to the customer that water is a
more limited and expensive commodity in the summer and less expensive
and more abundant in the winter.

The last cost-of-service study for the Phoenix Water Services Department was completed in 1987 and was used to develop the current rate
structure. Since 1987, the Union Hills treatment plant has doubled in capacity, Central Arizona Project water has become more expensive and a
rapidly growing proportion of the costs reflected in monthly water bills.
In addition the service area population has grown by over 6 percent. The
rate development process and the availability of data have also improved.
The primary recommendation by the Water Services Department was to
update the rate structure to reflect the cost of service to ensure that customers are paying their fair share for water service. All of the rate alternatives presented later in this report reflect a cost-of-service update.

### Equal Lifeline in All Months or Seasons

A major oversight in the development of the 1990 water rate structure was
how the larger summer lifeline would impact bills for some low water

users. As mentioned previously, the lifeline increases by a larger percentage than the volume charge between the medium and high season resulting in lower summer bills for some customers. This is in contrast to City of Phoenix cost of service. The Water Services Department recommended that this problem be fixed by making the lifeline equal in all of the months or rate seasons.

### Meter Maintenance Costs in the Volume Charge (Equal Service Charge for All Meter Sizes)

The 1990 water rate structure has a service charge that varies by meter size. This service charge contains the cost to read the meter, process water bills, and maintain the meter and service line. The meter reading and billing costs are considered to be equal for all meter sizes, whereas the meter and service line maintenance costs increase as meter size increases.

Recent studies by the City of Phoenix and the American Water Works Association suggest that the life of the meter is more impacted by the volume of flow through the meter than most other variables. In other words, the more volume a meter registers the more likely it is that it will require replacement. In addition, the data to determine the meter maintenance cost difference by meter size is difficult to obtain and may fluctuate annually.

By including the meter and service-line maintenance in the volume charge, those customers who use more water and wear out the meter faster will pay more. In addition, the service charge will be less for all customers and will provide even lower bills for low water use customers below the lifeline allocation. For these reasons, the Water Services Department believed it is more appropriate and equitable to include the meter and service-line maintenance in the volume charge and to have the service charge reflect only the cost to bill the customer and read the meter.

## Rate Structure Options

The Citizens' Water Rates Advisory Committee reviewed the concerns raised by customers and the improvements suggested by the Water Services Department. After their review, the CWRAC proposed some modifications to the current water rate structure, including a cost-of-service update option using the current rate structure.

The first option, which reflects the cost-of-service update of the current water rate structure, will be referred to as the Revised Current rate structure. This option maintains the current rate structure with a full cost-of-service update to reflect seasonal costs of service for fiscal year 1994–1995.

**Table 7-3.** Cost-of-Service Update of Current Water Rate Structure: Revised Current Option

| Consumption block in 100s of cubic feet | Monthly price per 100 cubic feet by season for all customers | | |
| --- | --- | --- | --- |
|  | Winter | Spring/fall | Summer |
| 0 to 6 | $0.00 | $0.00 | $0.00 |
| 7 to 10 | $0.62 | $1.06 | $0.00 |
| Over 10 | $0.62 | $1.06 | $1.63 |

| Monthly service charge | |
| --- | --- |
| Meter size (in inches) | Charge |
| ⅝ by ¾ | $4.78 |
| 1 | $6.30 |
| 1½ | $10.08 |
| 2 | $13.11 |
| 3 | $33.65 |
| 4 | $46.33 |
| 6 | $69.33 |

An 11.75 percent increase in revenues over the current rates is reflected in the cost-of-service update. If no new rate structure were considered, this would have been the recommended rate for fiscal year 1994–1995. Table 7-3 illustrates this option.

The second option, hereafter referred to as the Committee Recommendation, is the alternative selected by the CWRAC. This option maintains the three seasons but lowers the lifeline from the current level of 6 CCF in the low and medium seasons and 10 CCF in the high season, to 5 CCF in all seasons. In addition, the meter maintenance costs have been moved into the volume charge. This results in an equal service charge for all customers. As with the revised current option, this rate structure recovers 11.75 percent more revenue than the current rates and reflects a cost-of-service update. Table 7-4 illustrates the Committee Recommendation.

## Discussion of the Rate Options in Respect to Rate Objectives

Earlier in this report the rate objectives that served as the basis for the development of the current water rate structure were discussed. In order to make a determination of which of the two options is the most appro-

**Table 7-4.** Citizen's Water and Wastewater Rates Advisory
Committee Option: Committee Recommendation

| Consumption block in 100s of cubic feet | Monthly price per 100 cubic feet by season for all customers | | |
|---|---|---|---|
| | Winter | Spring/fall | Summer |
| 0 to 5 | $0.00 | $0.00 | $0.00 |
| Over 5 | $0.69 | $1.12 | $1.57 |

| Monthly service charge | |
|---|---|
| Meter size (in inches) | Charge |
| ⅝ by ¾ to 6 | $2.48 |

priate, both options are now discussed in relation to each individual rate
objective.

**Revenue Sufficiency**

*Revenue sufficiency* addresses whether the revenue flow from rates will be
sufficient and as stable as possible over time. Both rate structures are
designed to collect the same amount of revenue in the forecast year of fiscal year 1994–1995, so both can be said to be revenue-sufficient for that fiscal year.

Some financial experts argue that rate structures that place an emphasis on fixed charges are more revenue-sufficient and more stable than
those which emphasize volume charges. It is true that fixed charges are
guaranteed; however, so is a very large block of water consumption.
Businesses and households have a certain amount of water that is necessary to produce a product or to maintain a lifestyle. This block of consumption is virtually guaranteed and is thus as stable a revenue sources
as a fixed charge.

A method to estimate the revenue sufficiency and revenue stability of a
rate structure is to examine how a different rate structure would historically collect revenue. A revenue stability analysis of rate options using
data from 1986 to 1992 indicated that the committee option is more stable
in revenue collection during wet, cool weather when revenue stability is a
concern. This is in spite of the fact that the Revised Current option collects
twice as much in fixed charges. Several large public water utilities in the
United States have very low or no fixed charges in their rate structures
and have had no problems with revenue sufficiency related to the structures.

## Equity

*Equity* is described as the apportionment of costs according to the responsibility of costs. Both of the water rate structures are based on the concept of charging a volume charge that reflects the cost to provide service in different seasons or periods of the year. The only part of the rate that is not cost-of-service-based is the "lifeline" or consumption included in the service charge. This is a policy part of the rate structure to provide a benefit to lower water-using customers. The "lifeline" block of consumption provided to all customers specifically benefits low water use customers or customers with the majority of their consumption within the lifeline allotment.

Based on the preceding definition of equity, the most equitable rate could be defined as the rate structure that most closely causes the majority of the customers' monthly bill patterns to follow the monthly cost pattern of providing water service. This would be a rate structure that has the smallest service charge and the smallest amount of consumption in the service charge. The perfectly equitable rate structure would be one with no service charge and no "lifeline" allotment so that the average price paid for water service would be the same for all customers. Based on this criterion, the committee recommendation would be the more equitable since it has the lowest service charge and the lowest "lifeline" allotment.

The other problem with equity in both the current and revised rate structure is the improper price signal for users with low equal monthly consumption. As mentioned under the section Concerns with the Current Rate Structure, the larger summer lifeline causes some customer bills to decrease in the high summer months directly in contrast to the cost of service to supply water during this period.

## Efficiency

*Efficiency* is achieved when a rate structure encourages the use of water by customers in a way that minimizes social costs (financial, environmental, etc.). This is best achieved by pricing the water at the cost to replace the resource (the marginal price). Neither of the rate structure options is based strictly on marginal cost pricing; however, the higher summer and lower winter rate is a marginal price concept contained in both of the rate structure options.

If one agrees that the most efficient rate structure is one that comes closest to charging marginal costs to the most customers, then the most likely candidate would be the one with the smallest fixed charge, the lowest lifeline, and a volume charge most closely approximating the marginal cost. In addition, the most efficient rate structure would be the one where the largest percentage of customers pay the marginal cost of water. Because the lower fixed charge and the lower lifeline causes more customers to pay the marginal costs, the Committee Recommendation comes closer to achieving efficiency than the Revised Current option.

## Social Acceptability

This objective relates to whether the rate structure is acceptable to the public. This will depend upon the perceived equity of the rate structure and if the rates are understandable.

Based on experience with the current rate structure, customers seem to understand the reason for higher summer rates and most small users like the idea of consumption included with the service charge. However, some complaints have been registered by customers which were detailed previously in this report under the section, Concerns with the Current Rate Structure.

As mentioned under the section, Water Services Department Concerns and Recommendations, most of the problems can be addressed by making the lifeline equal in all months and reducing the service charge by placing meter maintenance cost into the volume charge. However, the large user complaint that the lifeline allocations result in higher rates for them can be addressed only by reducing or eliminating the lifeline. A reduction of the lifeline, if not followed by a lower service charge, will impact small users by shifting costs from larger to smaller water use customers. Social acceptability will depend on how large the impact is on small water users.

The Committee Recommendation attempts to address all of the concerns or complaints related to social acceptability of the current rate structure. This is accomplished by lowering the service charge, making the lifeline equal in all months, and reducing the lifeline to satisfy the concerns of large water use customers. The lowering of the service charge, in conjunction with reducing the lifeline, softens the impact on small water users. While this option softens the annual impact, the high or summer season impacts are the largest because of the reduction of the lifeline from 10 to 5 CCF. The value of the loss of 5 CCF to a small customer is an increase of $4.90 over their current bill of $5.43. Although purely on the basis of equity this customer will be paying a more fair share of the costs, the removal of the lifeline allocation to all customers will impact the small user the hardest. Even with this change, small single-family customers in Phoenix will still pay some of the lowest water bills in the Phoenix will still pay some of the lowest water bills in the Phoenix metropolitan area and Southwest. However, this was still the one area of the Committee Recommendation where social acceptability was of concern.

## Practical Feasibility

*Practical feasibility* relates to how feasible and practical the rate structure is to implement and administer. Both of the rate structures are implementable and capable of being administered. The Committee Recommendation is somewhat simpler to explain to the public because the amount of water in the lifeline does not change by season and the monthly fixed charge is the same for all meter sizes.

## Water Conservation

The Water Services Department defines *water conservation* as a beneficial reduction in water use. In other words, does water conservation meet the efficiency requirements described earlier?

Why worry about conservation as a rate objective? The Phoenix water system is designed and sized to meet maximum day and maximum hour demands in the summer months. Any sustainable effort to decrease these peaks will pay off in terms of energy and pumping costs and deferral of very expensive capacity expansions in the future. The intent of the design of the rate-structure options is to send the message that water is inexpensive in the winter and we have plenty of capacity, but in the summer supplies are more scarce and capacity is more limited. If customers receive this message through the water rates, the system will be more efficient in the long run. This is not unlike the approach used by electrical utilities to maximize the efficiency of their system.

An analysis of the long-term impact of the two options on water demand showed that the Committee Recommendation is slightly better at achieving the objective of water conservation by reducing annual water use an additional 0.40 percent, or slightly less than 1,000,000 gallons per day as compared to the Revised Current option. However, the Revised Current option is better at achieving conservation in the high season when demand reductions are the most cost-effective.

# Customer Bill Impacts

Any analysis of rate structure options is not complete without the most important analysis of which customers are impacted and by how much (customer impacts). Both of the proposed rate structure options reflect costs for fiscal year 1994–1995, which were projected to be 11.75 percent higher. In order to compare the options to the current rates, the current rates were increased 11.75 percent across the board. The comparisons of the across-the-board increase to the revised current option will reflect the change in customer bills due to the cost-of-service update. The comparison to the Committee Recommendation will reflect changes due to the rate-structure change and the cost-of-service update.

Water bill impacts on all customers are evaluated on small, medium, and large customers for a class. The small customers' consumption represents the 25th percentile or the consumption level, where 25 percent of the customers use less and 75 percent of the customers use more. The medium customer represents the 50th percentile, where 50 percent use more water and 50 percent use less water in a given month. Finally, the large customer represents the 75th percentile, where 25 percent of the customers in that group use more. The impacts illustrated are those impacts that will occur over and above the 11.75 percent revenue requirement. In other words, the rate structure options are compared to an across-the-board increase of 11.75 percent on the current rates to illustrate the impact of the rate structure change.

## Single-family Customers

On an annual basis, single-family customers are not impacted too severely by either of the rate-structure options. The small single-family customer realizes an annual bill decrease of $14.76 or 17.3 percent under the Revised Current rate structure and an annual bill decrease of $3.42 or 4.0 percent for the Committee Recommendation. The small single-family customer maintains usage below the current lifeline allotment in most months. Their decreased bills for the Revised Current option are mainly the result of the decrease in the ⅝-inch meter service charge based on the cost-of-service update. Their low use is also the reason for the large percentage increase in the high season for the Committee Recommendation. When the lifeline allotment is reduced from 10 to 5 CCF in the high season, small-use customers' monthly bills increase between four and five dollars. Although this may not be substantial on a dollar basis, on a percentage basis it is over a 70 percent increase. Figure 7-5 illustrates the change in monthly bills for the small single-family customer and indicates the large summer increases for the Committee Recommendation.

Monthly bill impacts are similar for medium and large single-family customers with the high season impacts slightly less than for the small customer. The Committee Recommendation increases high season bills by approximately 25 percent for large customers and 35 percent for medium customers. The larger the customer of any class, the less is the impact of lowering the lifeline allotment. This compares to a high season impact of the revised current rate structure of approximately 6 percent for the medium customer and 10 percent for the large single-family customer.

**Monthly Bill (Dollars)**

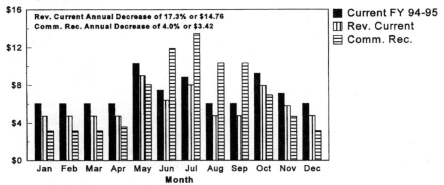

1) The percent difference between Current FY 1994-95 and the two alternatives is over and above the proposed 11.75 percent revenue requirement.
2) Demonstrates Single-family customers at the 25th Percentile of water usage.
3) Current FY 1994-95 = 11.75% across the board increase over current rates.
4) Rev. Current = Revised Current Rate reflects cost of service update: 6 ccf (10 ccf summer) lifeline, 3 seasons, unequal service charge.
5) Comm. Rec. = Committee Recommendation reflects cost of service update: 5 ccf lifeline, 3 seasons, equal service charge.
6) Rates do not include Water Environmental charge ( Current=$0.04/ccf , Proposed=$0.08/ccf).

**Figure 7-5.** Comparison of Monthly Water Bills: Current Rate Structure and Two Alternatives, Small Single-family Customer, Five-Eighth-Inch Meter

Generally, the low and medium season bills will decrease and the high season bills will increase for both options. The annual impact that results is a small decrease for the revised current rate structure and a small increase for the Committee Recommendation. Figures 7-6 and 7-7 illustrate the monthly water bills for the medium and large single-family customers, respectively.

### Multifamily Customers

Multifamily accounts are a type of water customer whose living units are served by a master meter. The majority of multifamily accounts are served by a 2-inch meter and the impacts discussed are for small, medium, and large 2-inch metered multifamily accounts.

Multifamily accounts with 2-inch meters are impacted by a small annual increase of 3.4 percent, 3.3 percent, and 1.9 percent for small, medium, and large customers, respectively, under the Revised Current option. The Committee Recommendation results in a small annual decrease of slightly less than 1.0 percent for small multifamily accounts and an increase of 1.6 and 1.8 percent for medium and large customers, respectively.

Both rate structure alternatives result in a decrease of nearly 25 percent in the low seasons for all sizes of 2-inch metered multifamily customers.

**Monthly Bill (Dollars)**

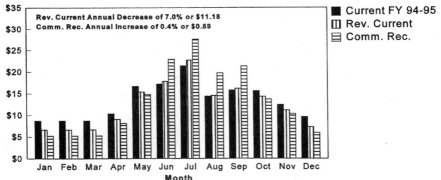

1) The percent difference between Current FY 1994-95 and the two alternatives is over and above the proposed 11.75 percent revenue requirement.
2) Demonstrates Single-family customers at the 50th Percentile of water usage.
3) Current FY 1994-95 = 11.75% across the board increase over current rates.
4) Rev. Current = Revised Current Rate reflects cost of service update:  6 ccf (10 ccf summer) lifeline, 3 seasons, unequal service charge.
5) Comm. Rec. = Committee Recommendation reflects cost of service update:  5 ccf lifeline, 3 seasons, equal service charge.
6) Rates do not include Water Environmental charge ( Current=$0.04/ccf , Proposed=$0.08/ccf).

**Figure 7-6.** Comparison of Monthly Water Bills: Current Rate Structure and Two Alternatives, Medium Single-family Customer, Five-Eighth-Inch Meter

**Monthly Bill (Dollars)**

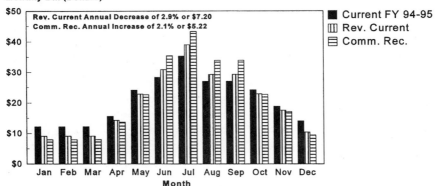

1) The percent difference between Current FY 1994-95 and the two alternatives is over and above the proposed 11.75 percent revenue requirement.
2) Demonstrates Single-family customers at the 75th Percentile of water usage.
3) Current FY 1994-95 = 11.75% across the board increase over current rates.
4) Rev. Current = Revised Current Rate reflects cost of service update:  6 ccf (10 ccf summer) lifeline, 3 seasons, unequal service charge.
5) Comm. Rec. = Committee Recommendation reflects cost of service update:  5 ccf lifeline, 3 seasons, equal service charge.
6) Rates do not include Water Environmental charge ( Current=$0.04/ccf , Proposed=$0.08/ccf).

**Figure 7-7.** Comparison of Monthly Water Bills: Current Rate Structure and Two Alternatives, Large Single-family Customer, Five-Eighth-Inch Meter

The revised current option results in an 18 percent increase in the high summer months while the committee recommendation is somewhat lower at nearly 13 percent. As a result the committee recommendation provides slightly lower annual increases than the revised current option. Illustrations of the monthly impacts for the different sized customers in this group were prepared but are not included in this case study.

### Commercial Customers

The impact on commercial customers is somewhat varied because of the large diversity in water use patterns in this group. The monthly impacts are similar to the multifamily customers; however, the annual increases are larger due to of a larger high season water demand by this class. Because the cost-of-service update increased the high season rate and actually decreased the low season rate, high summer use customers have the largest increase with both rate structures. The higher summer demand results in an annual increase of 11.7 percent, 8.1 percent, and 6.6 percent for the small, medium, and large customers under the Revised Current option. The annual impact is less under the Committee Recommendation because of the lower high season rate of $1.57 versus the $1.63 for the Revised Current option. The result is a diminished impact of a 3.7 percent decrease for the small customer, a 3.1 percent increase for the medium customer, and a 4.8 percent increase for the large customer under the Committee Recommendation. The monthly bill comparison graphs were prepared for CWRAC review but are not included in this case study.

### Irrigation Meters

As mentioned previously, the most impacted customers under either option are customers or accounts with high summer demands. The extreme of this case is a turf irrigation account, whose demand falls predominantly in the high season. An account with a 2-inch irrigation meter would experience a 13.5 percent annual increase for the Revised Current rate structure and a 9.8 percent increase for the Committee Recommendation.

### Summary of Customer Impacts

The impact on Phoenix Water Service customers from both options is the largest for customers whose consumption falls predominantly in the high summer months. Larger water use customers would experience summer bills approximately 13 percent higher for the Committee Recommendation option and 18 percent higher for the Revised Current option in the summer months. Conversely, the lower winter or low season rates are more attractive on an annual basis for customers with large winter demands. This is due to the cost-of-service update which indicates that the current rates are too high in the low or winter season period and too low in the high or summer period based on the cost of providing water service.

The Committee Recommendation would have provided the least difference in impacts among different types of customers on an annual basis. Small customers would have experienced a small percentage decrease while larger customers would have experienced a small percentage increase, with the largest impact on customers with exclusively high summer water use. However, small customers would see large percentage increases in the summer or high season period. Although the monthly dollar impact will be between $4.00 and $7.00 per month, the percentage impact would have been as high as 70 percent for the smaller customers. This compares to large water use customers that would pay 13 percent more in the summer months for the committee recommended option. The large percentage impacts to small customers in the summer or high season was due to the reduction of the lifeline allotment from 10 to 5 CCF.

## Summary Analysis of Proposed Rate Structures

The water rate structure implemented in 1990 has been successful in meeting the revenue requirements and objectives set by the Citizens' Water Rate Advisory Committee. However, the experience with the rate structure has pointed to areas for potential improvement to address concerns by customers and the Water Services Department. The CWRAC reviewed several rate structure options in relation to rate objectives and recommended a structure with a lower service charge equal for all meter sizes, a reduction of the lifeline allotment to 5 CCF in all seasons or months, and a cost-of-service update. This is compared here to the current rate structure with a cost-of-service update. Both rate options impact customers with increased bills in the summer months and lower bills in the winter or low season months. This is due to the update of the cost-of-service data, which reflects higher summer costs and decreased low season costs.

The test of the best rate structure is which one best meets the community objectives. In most cases the Committee Recommendation met the rate objectives better than the Revised Current option with the exception of the large percentage increases to small customers in the summer months. The large summer increases are due to the reduction of the lifeline allocation from 10 to 5 CCF in the summer or high season period.

Although the annual impacts to small users are nearly neutral, the summer impacts were expected to cause complaints. If the summer impacts can be addressed, the Committee Recommendation alleviates most of the concerns raised about the current rate structure and results in a rate structure that better meets the rate objectives and reflects the cost of providing water to the majority of City of Phoenix customers.

## Reaction to Recommended Rate Structure

Although the City Council had asked staff and the Citizens' Committee to address the concerns of the Multihousing Association, the solutions presented were not well received. As was expected, the Council was principally concerned with the significant percentage impact of both the Revised Current and the Committee Recommendation on summer water bills of single-family customers. They were also concerned that reductions in the lifeline would be viewed as a lack of interest in the poor and elderly.

Even though water rates in Phoenix remain low, the water bill provides a monthly reminder to all 300,000 customers that the city is charging for water. This reminder seems to result in more calls to City Council offices and questions at the Council's district forums than many more significant topics. The lifeline had also gained a "larger-than-life" image of the City caring about water customers that the council did not want to tamper with. On the other side of the equation was staff and a relatively small organization representing generically unpopular apartment developers. The organization also happened to be represented by someone out of favor with the council. The Citizens' Advisory Committee had also lost most of its original influence with the council.

The committee's response to concerns about reducing or eliminating the lifeline was to target low-income customers and reduce their bills. Although questions were raised about the administration of the program, it would have the benefit of providing relief to low-income customers using more than the lifeline amount of water. This would have represented a significant expansion of two small Water Services Department efforts to help low-income customers. One of these provided bill relief on a one-time hardship basis and another provided water audits and retrofits to reduce water use. The targeted lifeline did not gain sufficient support to overcome the concerns over eliminating the lifeline for all customers. Even a citizens' advisory committee to the City department providing services to the poor objected to taking money from most customers and giving it to a few. There were also concerns raised that the low-income customers requesting the targeted lifeline would be poor money managers and water wasters.

Addressing the concern over high summer bills proved to be insurmountable. One city council member even proposed raising the lifeline amount to provide more relief to low water users. There was some discussion of allowing customers the option of an annual budget payment plan if they objected to the high summer bills. In addition to reversing the original intent of sending proper price signals to customers during the summer, the City's aging computer system made implementing a budget payment plan impractical.

After months of analysis and discussion the water rate structure change died. The proposals made sense technically but not politically. This time no one was able to make the case for the changes sufficiently compelling that the city council would take the risk of upsetting thousands of water customers and being viewed as uncaring.

## Conclusion

There is a difference between what is desirable technically and theoretically and what is possible politically. This difference is not a problem of ivory tower theoreticians or corrupt politicians. It is a problem of balance and implementation capability. Balance refers to a decision maker's evaluation of the potential benefits of a proposed change and the potential costs. Is it obvious that the benefits are sufficiently compelling to warrant a decision maker's acceptance of the costs? Implementation capability refers to the proponents' ability to gain acceptance that the benefits are more valuable than the costs. While good technical analysis and theoretical grounding are needed, they don't necessarily result in acceptance of a proposal. Success in implementation has more to do with understanding the values held by groups and individuals and with overcoming the desire of a group or individual to stop acceptance of a proposal.

## References

City of Phoenix. 1986. *Water Conservation Plan 1986*. City of Phoenix Water and Wastewater Department: Phoenix, AZ.

City of Phoenix Water and Wastewater Department. 1990. Description and Evaluation of Current/Adopted and Previous Water Rate Structure. City of Phoenix, AZ.

City of Phoenix Water Services Department. November 30, 1993. Description and Evaluation of the Current Rate Structure and Two Proposed Alternatives.

Planning and Management Consultants, Ltd. 1986. *Water Conservation Evaluation for the Phoenix Water Service Area*. Carbondale, IL.

# 8

# Trends in Revenues and Expenditures for Water and Sewer Services: Implications for Demand Management

**David H. Moreau**
*University of North Carolina at Chapel Hill*

Among the more important factors in managing urban water demand in a particular community are the revenue and expenditure patterns of the utility serving that community, how the utility is financing its capital outlays, and how its customer charges have changed over time due to purposes other than demand management. Policies that modify demand also change both revenues and expenditures, and those changes are not necessarily of the same magnitude. The extent to which a utility may be able or willing to change customer charges, either as a demand-management policy itself or to neutralize financial effects of other policies, may well depend on how fast those charges are being adjusted for other reasons.

Factors of this kind should appropriately be evaluated using information from the utility that serves a particular community for which demand-management policies are being considered. For the nation as a whole, revenue and expenditure patterns, financing strategies, and customer charges have changed significantly since 1970. In this chapter, data from the United States Bureau of the Census and supplementary sources are used to examine the nature of revenues and expenditures for these services, to report trends that have been observed over the period 1970 to 1992, and to discuss some of the probable causes of changes in those trends. Among the most important causes has been the movement by local governments to make utilities financially self-supporting enterprises. The extent to which that has been accomplished is examined by comparing actual relationships among revenues, expenditures, and debt to those derived from a simplified model.

Finances of both water and sewer services are included in this discussion. Policies that change water demand may also change demand for sewer service, and, in many if not most utilities, the connection is direct because water consumption is used as the basis for sewer charges.

Only water and sewer services owned by local governments in the United States are considered here. Included are those systems owned by municipalities, counties, townships, and special districts. Exclusion of investor-owned systems from this discussion does not suggest that they are unimportant. Quite the contrary, they hold a nonnegligible share of the market, and a comparative analysis of public and private systems would be quite useful. Choosing to focus on local government-owned utilities was guided by several factors, the most important of which is that they are by far the largest class of providers of these services. The federal and state governments have, at times, provided substantial financial assistance to local governments for these services, but their direct expenditures are less than 1 percent of all government expenditures for these services. Investor-owned water utilities are primary providers in a number of urban areas, but they currently serve far fewer customers than do local governments. Only a very small percentage of sewer customers are served by privately owned sewer systems. A second significant factor in choosing to focus on local governments is the availability of a long-term consistent data set from which the needed information could be extracted.

## Categories of Expenditures

Water supply and sewer utilities are like other enterprises—they spend money to build facilities and acquire equipment and to administer, operate, and maintain services and facilities. Because facilities and equipment have economic value over many years, outlays for them are referred to as capital expenditures. Expenditures necessary to operate and maintain facilities, provide customer services, and administer the programs, are lumped under the broad category of operating expenses.

Water and sewer services differ from many other services provided by local government primarily in the proportion of all expenditures that go to capital outlay. For instance, in 1992, capital outlay accounted for 11 percent of all local government expenditures for other than water and sewer services; for these services, capital outlay came to 36 percent of all expenditures.

A wide variety of capital facilities are typically required for these services in large urban demand centers. Supplies are typically taken from surface water reservoirs, transported over considerable distances through transmission mains or other conveyances, treated, stored in ground-level and elevated tanks, and distributed through a vast network of pipes. A reverse pattern of flows and facilities is required for wastewater. A large network of pipes collects wastewater from water users and conveys it to central treatment plants. From there it flows through outfall lines to be released to surface waters.

Operating expenses also cover a wide variety of categories. One large category is for the personnel (salaries, wages, and fringe benefits) required to operate and maintain the facilities, provide customer services, and administer the utilities. A second major category is electricity. Water is a heavy commodity, requiring large amounts of electrical energy to lift and move it. Chemicals, including coagulants, disinfectants, and pH controls, are another substantial category of expenses. Insurance, contractual services, supplies, and equipment not included in capital outlays must also fall under the general category of operating expenses.

## Trends in Revenues and Expenditures

A rather extensive national data set is available for the broader categories of capital outlay, operating expenses, and revenues. The United States Bureau of the Census' (USBC) *Annual Surveys of Local Government Finances* reports selected financial characteristics of all local governments. Revenue reported by USBC includes charges and fees paid by customers; it does not include funds received through grants or proceeds from issuance of debt. Expenses include payments for goods and services, salaries, wages, and fringe benefits.

Capital expenses include all expenditures for construction projects and major equipment items regardless of the source of funds to pay for them. Furthermore, those expenses are counted within the year that facilities are built or equipment is acquired. Projects financed in part by federal and state grants are included as local government expenditures. Projects financed by debt are counted as expenditures at the time they are built. Interest payments on debt are counted in a separate expense category. Payments for the portion of the debt service that is principal are not counted as an expenditure. To count payments on principal in debt service would amount to double-counting—once when the project is built, and again when the debt is paid. Expenditures do not include contributions to reserve funds from which future capital projects will be financed.

When capital outlays are accounted for in this manner, it is not possible to draw immediate conclusions about the financial self-sufficiency of utilities by making direct comparisons between revenues and expenditures in any year. Budgets for utilities owned by local governments are most frequently set up in categories that make that direct comparison possible, because they focus on revenues needed to maintain a positive cash flow during each year, using categories such as operating expenses, debt service (including both principal and interest), and contributions to reserves. By contrast, when debt is used to finance a portion of capital outlay, revenues in any year may be less than expenditures as accounted for by the Bureau of the Census even in a financially self-sufficient enterprise. That is because revenue needed to cover debt service for an outlay in that year is deferred to future years. Similarly, if a utility is relying heavily on contributions to capital reserve funds to finance future projects, its revenue needs may exceed expenditures in any year.

For water supplies, noncapital expenses for water supplies are separated into two categories, operating expenses and interest on debt. Interest expense for sewer systems is not reported separately, but it is not included in expenses for that specific service. It is lumped with interest expense for all local government services not otherwise separately reported. Therefore, operating expenses for sewer service can be obtained by subtracting capital outlay from total expenditures.

Slightly more complete and detailed information is available from USBC's *Census of Governments: Finances,* published at five-year intervals. In addition to the previous information, these reports include debt outstanding at the end of the year for water supplies and sewer systems.

Raw data from USBC can be used directly to calculate some useful within-year ratios, but before meaningful interpretations can be made of time-series data, several adjustments are necessary. One of the more basic is to eliminate effects of inflation, which over the period 1970 to 1992 varied from a maximum of about 17 percent in 1981 to slightly negative in 1984, averaging about 6.0 percent a year over the entire period. In this case, the adjustment was made using the Gross Domestic Product (GDP) price deflators for nondefense spending, converting all data to equivalent 1992 dollars.

Other indices of inflation, such as the Producer Price Index (PPI), could have been used, but they would not have materially altered observations and conclusions drawn from the data. Use of the PPI would lead to slightly different results, especially during the high inflation years of the early 1980s.

## Aggregate Trends

Results obtained using the GDP price deflator for local government water supplies are given in Figure 8-1. Total expenditures for water systems have grown steadily since 1970. Revenues from this service remained rather flat during the 1970s, resulting in an increase in the difference between revenues and expenditures. After 1980, however, revenues from water services began to rise rather sharply. During the 22-year period, capital outlay remained rather flat, with the sharpest increases coming during the period 1985–1987. Operating expenses for water services rose steadily throughout the period.

Expenditure patterns for sewerage services, shown in Figure 8-2, reflect the pronounced influence of the federal construction grants program. Capital outlays by local governments for sewer facilities were growing rapidly in the early 1970s before the 1972 amendments to the Federal Water Pollution Control Act (now known as the Clean Water Act). Huge increases in federal subsidies authorized by that legislation greatly accelerated local government spending from 1974 through 1980. Spending for

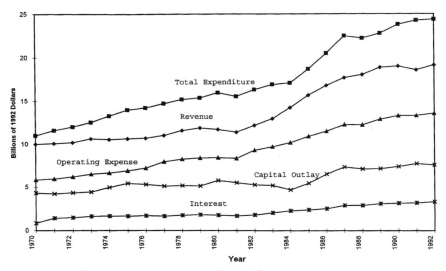

**Figure 8-1.** Water Supply Revenue and Expenditures

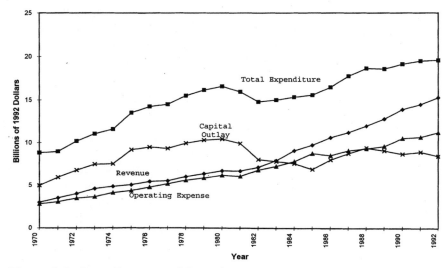

**Figure 8-2.** Sewer Revenue and Expenditures

capital projects increased from just about $5 billion a year in 1970 to over $10 billion in 1980. Capital outlay dropped to below $7 billion in 1985 as federal grants declined, but they stabilized in the range of $8.4 to $9.4 billion from 1987 to 1992.

One of the more noticeable trends in Figure 8-2 is the rapid increase in revenues in sewer services since 1980. From 1970 through 1982, revenues from sewer services were only slightly above operating expenses, and increases in expenditures were far outpacing increases in revenue. Since then, revenues have been increasing rapidly, narrowing the gap between revenue and expenditures.

Trends in Figures 8-1 and 8-2 are summarized in Table 8-1. Average annual increases in expenditures for the periods 1970–1980 and 1980–1992 are shown in addition to average increases over the entire period.

**Table 8-1.** Real Annual Increases in Revenues and Expenditures (in Millions of $1992 per Year)

|                  | 1970–1980 | 1980–1992 | 1970–1992 |
| ---------------- | --------- | --------- | --------- |
| Water            |           |           |           |
| Revenue          | 181       | 620       | 432       |
| Expenditures     | 484       | 703       | 609       |
| Sewer            |           |           |           |
| Revenue          | 356       | 712       | 559       |
| Expenditures     | 848       | 248       | 505       |

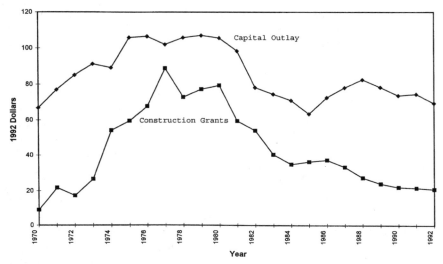

**Figure 8-3.** Capital Outlays for Sewer and Construction Grants per Housing Unit

As noted earlier, a major factor in the financing of sewer systems has been contributions from the construction grant program, as shown in Figure 8-3. During the 1970s, that program contributed 57 percent of local government capital outlays for sewer services; in the 1980s the contribution was 45 percent. As illustrated, however, the federal share has been in a general decline since its peak in 1977. Amendments to the Clean Water Act in 1987 directed a portion of those funds to capitalize state revolving funds. A long-standing intent of Congress has been to eliminate them entirely when the Clean Water Act is reauthorized.

### Household Trends

Aggregate increases of the kind given in Table 8-2 mask the fact that while revenues and expenditures have been increasing, so have the number of consumers of these services. Data from USBC's *Census of Housing* for 1970,

**Table 8-2.** Number of Housing Units Served by Central Systems (in Thousands)

|       | 1970   | 1980   | 1990   |
|-------|--------|--------|--------|
| Water | 55,293 | 72,528 | 86,069 |
| Sewer | 48,177 | 64,241 | 76,452 |

1980, and 1990 show that the number of housing units served by either publicly or privately owned water central systems increased 56 percent from 1970 to 1990. Housing units served by public sewer systems increased by 59 percent during that same period.

To estimate the average cost of serving one housing unit and to estimate the average amount paid by occupants of a housing unit, two other adjustments to the data are necessary. One adjustment must be made to account for the fact that the revenue data covers all customers, including commercial and industrial customers, not just residential customers. Expenditures also include costs to serve those other customers but losses in water distribution system. The other adjustment takes account of the fact that not all households are served by systems owned by local governments.

National estimates of water use are reported by the United States Geological Survey (USGS) can be used to make the first of these adjustments. For 1985 and 1990, USGS reported that domestic uses accounted for 57 to 57.5 percent of all publicly supplied water. Commercial use ranged from 15 to 16 percent, industrial use from 13 to 16 percent, and public use and losses from 11 to 14 percent. Estimates in prior years lumped domestic use with public uses and losses, but percentages reported for that lumped category ranged from 68 to 70 percent, making them consistent with the more recently reported figures. It is assumed for purposes of allocating revenue and expenses that residential use has been 57 percent and system losses have been 12 percent throughout the period. If revenue and expenses are also allocated in proportion to customer usage (losses being allocated in proportion to end use), then residential share is 65 percent (57 percent of 88 percent).

An argument can be made that all classes of customers do not impose the same unit cost on a system and not all classes pay the same average unit charge. Differences of that kind would introduce errors in the estimates, but only a small portion of total use is attributable to large customers where such differences could be important.

Errors from that source are probably smaller than those introduced by the other adjustment. A special survey of community water systems by the U.S. Environmental Protection Agency in 1986 reported frequency distributions of the sizes of systems by type of ownership. That data was used to estimate that 85 percent of customers were served by publicly owned systems, 15 percent by privately owned systems. In the absence of other regularly reported breakdowns between public and private systems, estimates of revenues and expenditures on a per-housing-unit basis were made by assigning 85 percent of housing units to publicly owned systems. The private sector provides a very small portion of centrally supplied sewer service, and, hence, all housing units on central systems were assumed to be served by systems owned by local governments.

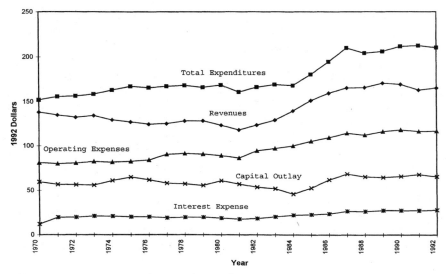

**Figure 8-4.** Revenues and Expenditures for Water Supply per Housing Unit

With these adjustments, trends in revenues and expenditures per residential unit are shown in Figures 8-4 and 8-5. There it may be noted that expenditures for water services were relatively flat during the period 1970 to 1980, and revenue per household unit actually declined during the period, reaching a low of $118 in 1981. From 1981 to 1987 both revenue and

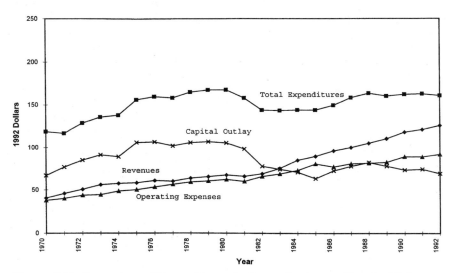

**Figure 8-5.** Revenues and Expenditures for Sewer Service per Housing Unit

expenditures rose sharply, and from 1987 through 1992 they remained relatively constant, revenue at about $165 and expenditures at about $210 a year.

Expenditures per household for sewer services followed a pattern similar to those of the aggregate patterns, showing sharp increases in the 1970s, dropping for the period 1982 through 1985, and rising to a new plateau since then, about $162 a year. Revenue derived from charges to customers grew steadily throughout the period, reaching $126 in 1992.

For some purposes it is important to track financial trends in water services separately from sewer services, but for the large proportion of customers who receive a single bill for both services based on metered water use, differences between the two may be immaterial. When revenue and expenditure data for the two services are combined, as shown in Figure 8-6, increases in revenue per housing unit are even more pronounced, jumping from $184 in 1981 to $283 in 1991, a 54 percent increase after adjusting for inflation. That increase followed the 1971–1981 period, during which the real increase was just under 5 percent.

## Debt

The Bureau of the Census, in its five-year *Census of Governments*, reports debt issued for water supply during the census year. End-of-year outstanding debt figures for both water supply and sewer utilities are also reported. For census years 1972–1987, debt issued to finance water supplies increased from 55 percent of capital outlay in 1972 up to 87 percent in

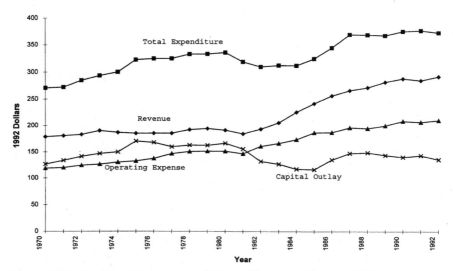

**Figure 8-6.** Combined Revenues and Expenditures per Housing Unit

1982 before dropping to 56 percent in 1987, an average of 68 percent during the period. The annual data in Figure 8-1 show that interest expenses have run consistently in the range of 11–13 percent of all expenditures. Similar data were not reported for sewerage, but the existence of construction grants for wastewater treatment plants made that percentage much lower.

Another indicator of the extent of debt financing is the ratio of capital outlay in any year to end-of-year outstanding debt. Those ratios are shown in Figure 8-7, where it can be seen that debt for water supply has been in the range of six to eight times current outlays. Local governments typically use repayment periods in the order of 20 years. With no growth and no inflation, the expected ratio would be 10 if 50 percent of capital outlay is financed by debt. As shown in the following section, when actual growth and inflation rates are accounted for, a ratio in the range 6 to 8 is consistent with debt financing in excess of 50 percent. Again, the much lower values for sewer systems reflect the influence of construction grants as a source of financing.

## Relationships in Self-Financed Utilities

For sewer services, the magnitude of the revenue-expenditure gap and contributions of construction grants make it obvious that for much of the

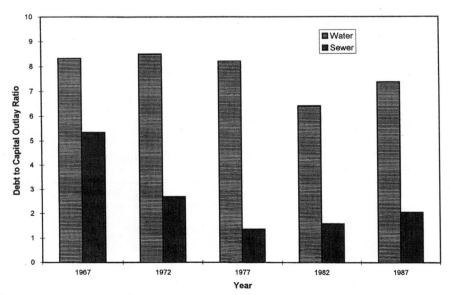

**Figure 8-7.** Ratios of End-of-Year Debt to Capital Outlay

past two decades, this service has not been fully financed by revenues from customer charges. Care should be exercised, however, before concluding that local government water supply utilities have not been self-sufficient or that sewer systems are still far from self-sufficiency.

What should be the relationship between revenues and expenditures and between capital outlay and debt in a utility that is fully self-financed by customer charges? To examine those questions, consider the relatively simple case of a utility that has experienced constant growth, constant per household costs, and constant financial factors over an extended period of time. In particular, consider the case where those factors were approximately equal to conditions faced by local governments in the United States in 1992. Let the number of households served grow at a constant rate of 2.0 percent a year. Let per household operating expenses for water and sewer services be initially $183 per year, and let capital outlays be initially $117 per year. Let inflation be constant at 3.0 percent a year. Assume debt is financed by 20-year serial bonds at an interest rate of 7.0 percent, and to avoid needless complications that add little to the analysis, let the interest rate on all series be constant regardless of time to maturity.

Under these conditions, expenses, revenues needed to satisfy cash flow requirements, end-of-year debt, interest expenses, and other factors can be calculated in a straightforward manner. Results are given in Table 8-3.

Revenues needed to balance cash flows are shown to drop from being equal to expenditures to being 85 percent of expenditures as the percentage of capital outlay that is debt financed increases from 0 to 100 percent. It can also be noted that the ratio of end-of-year debt to annual capital outlay increases from 0 to 9.2

This simple model illustrates that when debt financing is used, as it has been by local governments, revenues needed to balance cash flows are less than expenditures, even for a fully self-financed utility. Furthermore, ratios calculated using the simple model are quite similar to those observed for water supply services in the United States.

**Table 8-3.** Relationships in Self-Financed Utilities: Under Specified Conditions

| Percent of capital outlay financed by debt | Ratios of: needed revenue to expenditure | End-of-year debt to annual capital outlay |
| --- | --- | --- |
| 0 | 1.0 | 0 |
| 25 | 0.956 | 2.31 |
| 50 | 0.917 | 4.61 |
| 75 | 0.882 | 6.92 |
| 100 | 0.850 | 9.23 |

# Discussion of Trends

Although the model used in the preceding calculations overly simplifies some of the dynamic conditions experienced by local governments, it does provide a basis for interpreting some of the trends found in the data, particularly those affected by financing strategies. Other factors have also affected those trends, including trends in regulation of water and sewer utilities and increasing costs for personnel and other factors of production.

## Water Use

One variable that probably has not affected per household revenues and expenditures is patterns of water use. The USGS reports that per capita use of publicly supplied water has increased from 166 gallons per day (GPD) in 1970 to 195 GPD in 1990, pushed in part by increasing personal income. Over the same period, however, average household size declined from 3.14 to 2.63 as the number of households grew at a faster pace than the population. When estimates of water use are divided by the estimated number of households served by central systems, changes in use from 520 GPD in 1970 to 490 in 1990 are relatively small and subject to considerable reporting, rounding, and other errors. Increasing per capita usage has pushed up aggregate and per capita revenues and expenditures, but those increases are matched by an increase in the number of households served.

## Operating Costs

Steady increases in operating costs per household for both water and sewer services can be attributed to at least two factors: (1) more intense use of resources to achieve performance standards mandated in the Safe Drinking Water Act and the Clean Water Act; and (2) rising real costs for those resources. Slower growth in operating costs for water probably reflects the fact that water systems have been operated more intensely than sewer systems. Less improvement was needed in water system management than for sewer systems. One indicator of more intensive management in sewer systems was a 23 percent increase in city employees for this service from 1980 to 1990 while the increase in employees for water was a more modest 8.6 percent, about the same as the increase in production (Table 515, USBC, 1996).

Real costs for resources have also increased for several components of operating costs. Salaries and wages for personnel, which may account for as much as 40 percent of operating expenses, have increased in real terms. Fringe benefits for personnel have also risen sharply. The consumer price index for electricity, another significant component of operating expenses, increased faster than the index for all consumer items, especially in the early 1970s and from 1980 to 1985.

## Influence of Financing Practices

Among the more interesting trends is sewer revenue relative to expenditures. Throughout the 1970s revenues remained only slightly above operating costs, despite the fact that spending on capital outlay was much greater than operating expense. At the beginning of the decade, before the great flow of funds from federal construction grants, local governments were spending $1.63 for capital outlay for every $1 on operating expense. At the end of the decade, they were spending $1.69 on capital for every $1 on operations. With customer revenues only 1.08 times operating expenses, it is clear that very little capital outlay was being financed by that source.

Despite the pronounced effect of the construction grant program, local governments still had to find substantial funds for capital outlay from their own sources. Because revenues from customer charges were insufficient, some other source was obviously being used to fund the local government share. Some other source of funds was obviously being used to fund the local government share. Regularly reported data are not available to estimate contributions from alternative sources, but it was commonplace for local governments to finance sewer improvements through general obligation bonds that were repaid from property tax revenues.

Two factors shifted that situation in the 1980s. First, as the flow of federal construction grant funds declined, local government capital outlays for sewer systems declined. That occurred as operating costs continued their steady rise. In 1984, for the first time in at least several decades, operating expenses exceeded capital outlay. Second, a widespread movement among local governments shifted water, sewer, and other publicly owned utilities to so-called "enterprise funds"—operating them in a financially self-sufficient manner and making it easier to carry over funds from one budget year to the next. That change was part of a broader movement within government to relieve some of the burden on property taxes by placing more of it on user charges. Because water and sewer services are revenue producers, they were among the first to experience this change.

The rapid rise in sewer revenues from 1982 through 1992 are indicative of a utility moving toward financial self-sufficiency. As federal funds declined, local governments began picking up a larger share of costs to expand and upgrade facilities. Those costs were passed on to customers in the form of higher rates. In 1980, revenues covered only 40 percent of expenditures; in 1992 they covered 78 percent of expenditures.

It is clear from trends in water revenues relative to expenses that water systems have been operated much closer to financially self-sufficient enterprises than have sewer services. In the early 1970s revenues from water were 85 percent of expenditures, consistent with about what would be expected by a financially self-sufficient utility. That percentage dropped to about 75 percent in the period 1977–1982, after which it began to rise, reaching 79 percent in 1992. At least part of that change can be attributed to increased reliance on debt financing during the 1970s, as discussed earlier. With increased debt, the need for revenue to cover expenditures as they occurred decreased. With high interest rates in the early 1980s, reliance on debt financing decreased sharply, and revenues had to rise to pay a greater portion of capital outlays from current funds.

The sharp rise in water revenues in the mid-1980s reflects that change in financing practice, but it probably also shows the same kind of move toward financial self-sufficiency that was more evident with sewer services.

## Summary and Conclusions

Three very significant factors affected trends in local government revenue and expenditures for water supply and sewer services over the period 1970–1992. First, throughout the period, operating expenses per unit of housing served rose steadily for each service. Operating expenses for the combined services rose by 76 percent after adjusting for inflation, an increase that can be attributed to more intensive management necessary to satisfy performance standards established pursuant to the Safe Drinking Water Act and the Clean Water Act. Rising personnel cost also contributed to that increase.

Second, the federal construction grant program had a profound effect on trends in capital outlay by local governments as those funds increased dramatically in the 1970s and declined in the 1980s. It also permitted local governments to maintain low customer charges during the 1970s.

Third, the movement toward financing based primarily on revenue from customer charges, in combination with declining federal assistance, led to rapid increases in consumer rates from 1980 through the end of the period.

Trends in the ratio of revenue to expenditures are nearing that which would be expected from utilities that are fully financed by customer charges. The gap that existed between actual revenues and those needed to fully finance these services has been substantially narrowed, resulting in increases in customer rates since 1992 that should have moderated, tracking more closely increases in expenditures. That can be verified only as additional data become available.

Despite increases from 1980 to 1992, customer charges for water and sewer services remained low relative to household income. At about $280 per housing unit in 1992, those charges were less than 1 percent of household income. At that level, the average household is not likely to react strongly to changes in prices for these services, especially for nondiscretionary uses of water.

These observations have several implications for management of demand for urban water. First, if households are adjusting to price changes, they have an incentive to do so since 1980. Even though prices have been increased for reasons other than managing demand, they have risen. How those increases have affected use is addressed in other chapters of this book.

Second, urban water and sewer utilities are highly capital intensive, although as shown by the data presented here, that intensity has declined somewhat in recent years. The implication is that reductions in demand that reduce the need for capital can have a significant impact on long-term revenue needs. The limitation is, of course, that those are long-term benefits. In the near-term, because capital expenditures are largely fixed costs, reductions in demand could lead to revenue losses that exceed reductions in expenses. The magnitude of that shortfall will depend on the extent to which capital expenses are being recovered by demand-dependent elements of the rate structure. In communities where that may occur, some of those impacts can be mitigated through redesign of the rate structure, either by increasing commodity charges or by recovering a greater portion of capital expenses through fixed charges and special capital fees.

## References

United States Bureau of the Census. 1972–92. *Annual Surveys of Local Government Finances.*

United States Bureau of the Census. 1996. *Statistical Abstract of the United States,* United States Government Printing Office, Washington, DC.

Solley, W. B., R. R. Pierce, and H. A. Perlman. 1993. *Estimated Use of Water in the United States,* Circular 1081, United States Geological Survey, Washington, DC.

# 9

# Demand Management Planning Methods

**Eva M. Opitz**
*Planning and Management
Consultants, Ltd.*

**Benedykt Dziegielewski**
*Southern Illinois University
at Carbondale*

As described in Chapter 1, water demand management (or conservation) is increasingly recognized as an option for balancing water supply and demand needs. One of the critical questions in water conservation planning is, among the many water-management tools available to the policy maker for providing water to customers (including supply enhancement and demand management), what level of conservation is optimum? From another perspective, how much investment in water conservation measures can be justified? The answers to these questions will depend on many factors, especially on the degree to which reduction of water demands will lead to the alteration of the long-term plan for acquiring new supplies and expanding or improving the existing water supply system. In order for demand management options to be given the same level of consideration as supply alternatives, the process for selecting and implementing programs must include rigorous consideration of potential effects, including customer acceptance, water savings, benefits, and costs.

In this chapter, a procedure is outlined for a systematic analysis of water conservation alternatives. The procedure consists of series of steps which, if executed, will permit the water resource planner to formulate viable conservation alternatives and decide upon the optimal level of water conservation in the long-term water-management plan. A careful examination of the following should be undertaken before decisions are made to implement a demand management program and incorporate the program into the long-term water-management plan: (1) conservation goals, (2) applicability and technical feasibility, (3) social acceptability, (4) implementation conditions, (5) potential water savings, and (6) potential benefits and costs. These issues, which comprise the steps in the procedure, are addressed in the following sections.

## Establish Program Goals

In developing a water demand–management program, the first issue that must be considered is the overall purpose of the program. In general, a long-term objective of a demand management program may be to achieve and maintain water use efficiency within an agency's water service area. However, while designing and implementing various water demand management programs, it is also important to translate the overall objective into specific goals. Specific goals of the demand management program should be based on an understanding of the local conditions of supply and demand. These goals will allow a program planner to determine those components of water demand that need to be monitored or modified by the demand management program. Assuming that there is a problem in the balance of water supply and demand or in the water/wastewater distribution or treatment system, several questions can be posed to assist in the development of the water conservation program goals:

1. Is there a short-term (e.g., drought-related, source contamination, or other emergency condition) or long-term (e.g., inadequacy of long-term supplies or storage capacity) water supply problem?

2. Is there a distribution system or capacity problem (e.g., excessive sewer flows, water/wastewater treatment plant capacities)?

3. Is the problem localized (e.g., capacity problems of a single water or wastewater treatment plant) or systemwide?

4. Is the problem seasonal in nature (e.g., summer demands, maximum daily demands, or average annual demands)?

Even if the water utility is not facing a supply, treatment, or distribution system problem, there may be utility standards or state/local mandates for water use efficiency or conservation. By assessing the need for and objectives of, a water conservation program, goals can be set and water conservation measures can be screened with respect to meeting those goals. Given the local water supply and demand issues, the conservation goals may be quantified (e.g., reduction of 10 percent of peak-daily demands or reduction of 2 MGD of average-daily demands). The establishment of conservation goals sets the stage for an analysis of potential conservation measures.

# Determine Applicability and Feasibility

The purpose of this step is to present methods for an initial screening of specific water conservation measures (i.e., practices, techniques, devices) for potential implementation and further evaluation. The recognition of local problems and the knowledge of water use in the service area will facilitate the screening process. The initial screening of a large number of conservation practices will ensure that the most promising conservation measures for a given water service area are not overlooked, thus enhancing the overall efficiency of a conservation program. At the present time, the accumulated experience of water agencies in the United States and elsewhere can provide a conservation planner with over 100 conservation measures (or variants thereof). Therefore, the initial screening of such a large number of conservation measures will assist the planner in selecting and evaluating only those measures that are feasible, acceptable by water agency customers, and effective in reducing water use. Typically, the screening process relies on information about the effects of various measures that are found in the literature or derived from the experience of other water supply agencies. For most measures, such information will be adequate only for the purpose of the initial screening. Economic analyses of conservation alternatives (addressed later in this chapter) as well as an empirical analysis of pilot programs (see Chapter 10, Demand Management Program Evaluation Methods) will provide more reliable and complete data for making final decisions about whether various conservation programs should be implemented.

## A Library of Conservation Measures

There are a large number of potential water conservation measures that may be used to reduce the use or loss of water in a particular water service area. An extensive listing of such measures should be obtained or developed and used as a starting point in selecting and evaluating conservation measures.

Tables 9-1 and 9-2 present two alternate typologies of long-term water conservation measures. The discussion here focuses on long-term water conservation measures rather than short-term (or drought-management measures). Table 9-1 groups conservation measures by types of water use in urban areas. Table 9-2 categorizes various measures according to the method of their implementation.

### Screening Tests

An initial screening of conservation measures can be conducted by assessing their *applicability* and *technical feasibility*. Screening for applicability defines those conservation measures applicable to water uses that take place in the water service area. For example, a conservation measure targeting landscaping practices may not be applicable if there is not a measurable seasonal (or outdoor) water use component. The conservation measure should also address the water supply problem that was the impetus of the conservation program. If the objective for the water conservation program is to reduce demands on an overburdened wastewater treatment plant, implementing lawn-sprinkling restrictions would have little or no effect on this goal.

The test of *applicability* should also define conservation practices that have been implemented in the water service area. If a measure is already implemented for a portion of the water service area, or for some (but not all) water uses in the area, an applicable measure is one that would apply to that portion of the water service area or to those water uses not already affected.

Measures are deemed *technically feasible* if, upon implementation, they result in a measurable reduction in the quantity of water used. Engineering analysis and reports of field studies may be used to establish apparent technical feasibility. In some cases, preliminary field tests of specific measures may be used to establish technical feasibility. For example, field tests of devices for reducing toilet-flushing volumes (plastic dams) may reveal that, for some types of toilets, flushing efficiency is so reduced that double-flushing occurs. If this leads to the determination that the devices would not achieve any measurable reduction in water use, the measure would be deemed technically infeasible.

# Determine Social Acceptability

Social acceptability is essential for the successful implementation of a water conservation measure. Therefore, an assessment of social accept-

**Table 9-1.** Typical Long-Term Water Conservation Measures by Water Use Type

General
    Public information
    Metering
    Pricing policies
        Uniform commodity rates
        Increasing block rate
        Seasonal rates

    In-school education
    Pressure reduction
    Leak detection and repair
    System rehabilitation

Interior domestic use
    Toilets
        Early closure flapper valve
        Toilet leak detection and repair
        Ultra-low-flush toilets
        Toilet displacement bags
        Dual flush devices
        Fill-cycle regulator

    Bathroom and kitchen faucets
        Low-flow faucets
        Faucet aerators
        Faucet washer
    Dishwashers
        Water efficient dishwasher
    Washing machine

    Showers
        Low-flow showerheads
        Shower-flow restrictors
        Shut-off valves

        Water efficient vertical axis
        Horizontal axis
    Air conditioning
        Air-cooled systems

    Urinals
        Ultra-low flush urinals
        Waterless urinals
        Valve retrofit

        Water efficient evaporative coolers
    Water treatment devices
        Water efficient reverse osmosis filters
        Water efficient water softeners

Landscape irrigation/design
    Efficient landscape design
    Water efficient plant material
    Efficient irrigation systems
    Garden hose timer
    Soaker hoses
    Greywater systems
    Xeriscape incentives
    Soil moisture sensors
    Turf watering literature

    Garden hose timer
    Scheduled irrigation
    Tensiometers
    Reduction or limitation of high water
     use plant materials such as turf
    Cisterns
    Turf reduction
    Rain sensors
    Peak management scheduling

Other outdoor use
    Hose control nozzles
    Water recycling/recirculating
     systems

    Swimming pool/spa covers
    Water efficient misting systems

Commercial/industrial use
    Recirculation of cooling water
    Reuse of treated wastewater
    Reduce "blowdown" on evaporative
     coolers, boilers, cooling towers

    Reuse of cooling and process water
    Process modification
    Equipment metering for leak detection

SOURCES: Maddaus, W. O., (1987); Seattle Water Department (1992); Barakat & Chamberlin, Inc. (1994), (1995).

**Table 9-2.** Typical Long-Term Conservation Measures by Mode of Implementation

| Education | Regulations | Management |
|---|---|---|
| Direct mail | State and local codes and ordinances | Leak detection and repair |
| Pamphlets, bill inserts, newsletters | Plumbing codes for new structures | Metering |
| Mass media | Landscape ordinances | Pressure reduction |
| Radio, TV, newspaper | Restrictions | Water reuse/recycling/recirculation |
| Personal contact | Rationing | Pricing policies |
| Speaker programs, customer assistance hotlines | (1) Fixed allocation | (1) Marginal cost pricing |
| | (2) Variable percent plan | (2) Increasing block rate |
| Special events | (3) Per capita use | (3) Seasonal rates |
| School programs, exhibits | (4) Prior use basis | (4) Summer surcharge |
| | | (5) Excess use charge |
| | | Tax incentives, subsidies, and rebates |
| | | Voluntary implementation of water-saving devices |
| | | On-site water audits |

SOURCE: Baumann et al. (1980).

ability is necessary in the evaluation of any water conservation measure, since it determines the probable response of the community to the proposed measure. This, in turn, provides practical information for the calculation of the expected level of coverage (or market penetration) of a conservation measure (see later section on estimating potential water savings). Measures are socially acceptable if they would be adopted by the community in which they are proposed. Unlike technical feasibility or economic feasibility, however, only rarely can the social acceptability of a given water conservation measure be predicted with a high level of certainty. The goal of this assessment is to increase the quality of the judgment regarding the probable response of various sectors of the community to a proposed measure.

Oftentimes, the social acceptability of conservation measures is determined based upon the intuitive considerations of water conservation planners or utility managers. This determination may be founded upon the successful application of the conservation measure in other communities. However, the successful application of a conservation measure in other communities does not ensure its success in all communities. Alternately, the failure of a specific conservation measure in one community does not ensure failure in all locations. Differences in socioeconomic characteristics, attitudes, and water use behaviors may preclude the same level of public acceptance as observed elsewhere.

The judgment of social acceptability can be assessed by initiating a survey research program that addresses two sectors of the community: (1) community leaders and interest groups and (2) the general public. Input to the water conservation process can be elicited through workshops, personal interviews with community leaders and interest groups, or surveys of the general public.

Because it is nearly impossible and not cost-effective to obtain information from all customers within a water utility service area, the survey addresses one or more samples of customers. Sampling has many advantages over a complete enumeration (or inventory) of the population under study. These advantages include reduced cost, greater speed of obtaining information, and a greater scope of information that can be obtained.

In order to develop a survey and sampling approach to meet the objectives of measuring social acceptability, these six steps are necessary:

1. Review objectives and define target population
2. Define data to be collected
3. Select method of data collection
4. Design and implement sampling plan
   *a.* Define sample population and sample frame
   *b.* Determine sample size
   *c.* Select a sampling method
   *d.* Implement sampling plan
5. Design and implement survey questionnaires
6. Data compilation and analysis

Each of these steps is described in the following sections.

## Review Objectives and Define Target Population

The obvious starting point for measuring social acceptability is to review the study objectives. For example, the following may be goals of the study:

- *Goal:* Characterize the potential acceptability of specific conservation measures of the residential population.

  *Target:* All residential customers in service area.

- *Goal:* Characterize the potential acceptability of commercial/industrial water conservation programs.

*Target:* All nonresidential customers in service area.

- *Goal:* Assess the general water conservation attitudes and behaviors of customers.
*Target:* All customers in service area.

- *Goal:* Assess the potential acceptability of a program that will target water use efficiency in large landscape customers (greater than 2 acres).
*Target:* Large landscape customers such as schools, hospitals, city parks, cemeteries, golf courses, and other select commercial/industrial accounts.

- *Goal:* Determine the likelihood of adoption of various conservation programs.
*Target:* Community leaders and interest groups.

The sampling and survey approach will be a function of the specific study objectives. Therefore, it is necessary that the objectives be defined as clearly as possible. In some studies, a number of different sampling and survey approaches may be necessary to meet the desired objectives. These objectives will indirectly define the desired precision in the measurements as well as determine the customer classes and types of water use that need to be evaluated in the study.

The target population of the study can be very broad (covering all customer classes) or it can be very narrow (covering only one customer class). Obviously, the scope of the study will directly affect the costs and timing of the effort. For an assessment of social acceptability of various potential conservation programs, it may be appropriate to sample and analyze customer groups. Most conservation programs target specific customer groups because of the similarity in end uses of water. The aggregation of water customers into groups also provides two advantages for the sampling process: (1) by decreasing the number of populations to be sampled, the number of different survey instruments and sample analyses can be held to a manageable number, and (2) by creating a smaller number of relatively uniform groups, the overall number of customers that need to be sampled can be reduced.

## Define Data to Be Collected

For the purpose of measuring social acceptability of conservation measures/programs, there are three types of detailed information that can be sought to characterize the customer base:

1. Situational characteristics of customers (e.g., residential: lot size, housing type, irrigable area, landscape types, household size, household

income; nonresidential: type of business/service, number of employees, etc.)

2. General attitudes and behavior toward water use such as water-using habits, perceptions regarding water supply issues (e.g., need for additional water supply or perception of potential shortages), or attitudes toward conservation (e.g., need for conservation, equity issues, efficacy issues)

3. Likely adoption of specific water conservation program initiative (e.g., likelihood of participating in a ultra-low-flush rebate program or a home water audit)

The data to be collected during the social acceptability study need to be defined precisely. As a direct function of the study objectives, the selected data to be measured will have a direct impact on the survey and sampling approaches.

## Selecting a Method of Data Collection

Generally, there are three types of survey methods: (1) mail surveys; (2) telephone surveys; and (3) personal interviews (which may include workshops, focus groups, or field surveys). Each of the methods has its comparative strengths and weaknesses with regard to administrative costs, data quality, and obtaining a representative sample. Table 9-3 compares the three survey methods qualitatively. The advantages and disadvantages of the survey methods are then briefly described.

Mail surveys tend to be the least costly of the three approaches. For this reason, mail surveys can be used to obtain relatively large samples. Because the surveys are completed at the respondents' convenience and because the questions can be reread at their discretion, more complex and detailed questions can be asked. However, the response rates to mail questionnaires can be very low if not followed up with reminders (postcards or additional questionnaires). But, follow-up reminders will add to the cost of administering the survey. Mail surveys also lack the control of other survey methods, including clarification of procedures, question ordering, or selection of respondents (e.g., head of household). Because response rates tend to be lower for mail questionnaires, there is a greater chance of nonresponse bias (i.e., nonrespondents having significantly different characteristics than respondents).

Telephone surveys tend to have the fastest turnaround time. Although they tend to be more costly than mail surveys, they are less costly than personal interviews. Telephone surveys can be used to target the appro-

**Table 9-3.** Comparison of Mail, Face-to-Face, and Telephone Surveys

| Factor | Mail | Face-to-face | Telephone |
|---|---|---|---|
| Administration | | | |
| 1. Cost | 1 | 4 | 2 |
| 2. Personnel requirements: interviewers | n/a | 4 | 3 |
| 3. Personnel requirements: supervision | 2 | 3 | 4 |
| 4. Time for implementation | 4 | 4 | 1 |
| Sample | | | |
| 5. Sample coverage | 3 | 1 | 1 |
| 6. Response rate—general public | 4 | 2 | 2 |
| 7. Refusal rate | unknown | 3 | 3 |
| 8. Noncontact/nonaccessibility | 2 | 3 | 2 |
| 9. Ability to obtain response from elite population | 4 | 1 | 2 |
| 10. Respondent within household | 4 | 1 | 1 |
| 11. Sampling special subpopulation | 4 | 2 | 2 |
| Data quality | | | |
| 12. Interview control | n/a | 3 | 1 |
|     *a.* Control consultation | 4 | 1 | 1 |
| 13. Obtaining socially desirable responses | 1 | 4 | 3 |
| 14. Item nonresponse | 3 | 2 | 3 |
| 15. Impact of questionnaire length on response | 3 | 1 | 2 |
| 16. Confidentiality | 4 | 4 | 3 |
| 17. Ability to ask sensitive questions | 2 | 1 | 2 |
| 18. Ask complex questions | 3 | 1 | 3 |
|     *a.* Ability to clarify | 4 | 1 | 2 |
|     *b.* Use of visual aids | 3 | 1 | 4 |
| 19. Use of open-ended questions | 4 | 1 | 2 |
|     *a.* Ability to probe | 4 | 1 | 2 |

SOURCE: Developed from Fowler (1989) and Frey (1988).

Key: 1 = major advantage.
    2 = minor advantage.
    3 = minor disadvantage.
    4 = major disadvantage.
    n/a = not applicable.

priate population (e.g., heads of households, program participants). This method also allows more in-depth, probing questions and will allow clarification of responses. However, telephone survey methods can also be subject to nonresponse bias from households without telephones or households with unlisted numbers. When sampling from telephone directories or other listings, unlisted telephone numbers can cause significant nonresponse bias. However, random-digit dialing methods, typically used by marketing research firms, can overcome the problem of unlisted telephones and can also help reduce sampling bias. The length of telephone interviews is also limited and therefore may restrict the number of questions and may also limit the complexity of questions. It should be noted that the interviewer's contact with the respondent may also lead to potential biases.

Personal interviews (field surveys) tend to be the most costly survey method. However, there is much greater control over the selection of respondents, and response rates tend to be higher than for the other methods. The time respondents are willing to devote to personal interviews tends to be longer than with telephone interviews. Therefore, more complex questions can be asked and the interviewer can provide more clarification. As with telephone interviews, the personal contact between the interviewer and the respondent may lead to response bias. With personal interviews conducted at a respondent's residence (or business), situational data can be obtained without even posing questions (type of residence, type of yard, size of house). Personal interviews tend to be practical only for small sample sizes without large geographic constraints.

The types of data to be measured in the study will, in some cases, drive the survey method to be used. For example, if the desired measured parameters include the flow rates of toilet, showerheads, and sprinkling systems, this will require on-site measurements and will preclude conducting telephone and/or mail surveys.

## Design and Implement
## Sampling Plan

The development and implementation of the sampling plan will be one of the most critical aspects of the measurement of social acceptability. The sampling plan and its success will directly affect the representativeness of the data obtained from the survey. Each component of the sampling plan is described in the following sections.

**Define Sample Population.** The sampled population should coincide with the target population. The latter denotes the aggregate from which the sample is chosen. The analyst must decide what group of customers is to be investigated. For example, it is necessary to determine if the results obtained from the sample should apply to all residential customers or to all single-family homes. A precise definition of the sampled population will allow the field personnel to decide whether a doubtful case belongs to the population. In cases where the sampled population is smaller than the target population, the conclusions apply only to the sampled population. In order to apply the findings to the target population, supplemental information will be required on the nature of the differences between the target and the sample population.

**Determine Sample Size.**   For determining social acceptability of a conservation measure or program, the typical desired information is a *proportion* of respondents who provided a given response (e.g., 60 percent of households have low-flow showerheads; 40 percent of households do not). This is referred to as *sampling for proportional data.*

When determining a sample size requirement, it should be noted that a single sample size estimate for an entire survey cannot be known; rather, a sample size requirement can be calculated for only one survey question at a time. For a desired level of precision, the sample size requirement may substantially vary for various questions of the survey. To estimate the sample size requirement for proportions, the margin of error, level of confidence, and the estimated proportion of the measured variable (e.g., 50 percent/50 percent or 80 percent/20 percent) must be specified. It should be noted that at a 50 percent proportion, the sample size requirement is at the maximum level. As the expected proportion moves away from a 50/50 split, the sample size requirement is reduced for a given level of error and confidence. It is for this reason that most sample size estimates for proportions are based upon a 50/50 percent split for a given level of error and confidence.

Table 9-4 shows the relative error (+ or −) that can be expected from different sample sizes and different proportional splits on responses. For example, if you want to determine the sample size for a variable that is expected to have a 50/50 split (50 percent homes with low-flow showerhead, 50 percent homes without) with ±4 percent error, then you would need a sample of about 500 from each group you wish to sample. It should be noted that the estimates of sample size requirements only account for sampling error and do not account for nonresponse error, data errors, or measurement errors. Therefore, when selecting a sample for a survey, the sample sizes should be larger than what is required for desired levels of precision. For example, when conducting sampling for a mail survey, the nonresponse rate must be taken into consideration. When conducting a telephone survey, the working phone rate, the incidence rate (of reaching targeted population), and the potential completion rate need to be taken into consideration.

**Select a Sampling Method.**   Once a sample size has been designated, the analyst will want to develop a procedure for selecting the desired sample. There are two main approaches to sampling: probability sampling and nonprobability sampling. *Probability sampling* refers to any sampling

**Table 9-4.** Relative Error (+ or –) for Variability Due to Sampling for Proportional Data (Assuming 95 Percent Confidence)

| Sample size of each group | 5/95 | Percent of sample with characteristic | | | |
|---|---|---|---|---|---|
| | | 10/90 | 20/80 | 30/70 | 50/50 |
| 100 | 4 | 6 | 8 | 9 | 10 |
| 200 | 3 | 4 | 6 | 6 | 7 |
| 300 | 3 | 3 | 5 | 5 | 6 |
| 500 | 2 | 3 | 4 | 4 | 4 |
| 1,000 | 1 | 2 | 3 | 3 | 3 |
| 1,500 | 1 | 2 | 2 | 2 | 2 |

SOURCE: Fowler (1989) and Planning and Management Consultants, Ltd. (1994).

procedure that relies on random selection and is amenable to the application of sampling theory to validate the measurements obtained through sampling. This requires that within the sampled population one is able to define a set of distinct samples (where each sample consists of sampling units) with known and equal probabilities of being selected. One of these samples is then selected through a random process. *Nonprobability sampling* refers to sampling procedures that do not include the element of random selection. Assuming that the information is available to construct and implement a probability-based sample, this method is usually preferred over nonprobability samples and will likely result in a more representative sample.

**Implement Sampling Plan.** In some cases, samples can be selected and recorded prior to obtaining the data (e.g., the development of mailing lists). In other cases, sample selection is performed concurrently with data collection. The implementation of the sampling plan should be monitored closely. If response rates are less than originally anticipated, adjustments may have to be made in the sampling approach. Caution should be used in deviating from the approach of the original sampling plan.

## Design and Implement Survey Questionnaires

The type of survey method chosen (i.e., mail, telephone, personal interview/field survey) will have an impact on the development of the survey questionnaire (instrument). The design of the survey instrument is crucial to the successful measurement of social acceptability. Not only is it important what is asked, but it is important how it is asked. There should be a purpose for every question, and it should fit into the overall purpose of the survey. Careful attention must be given to all aspects of survey development. The major areas of concern are question formulation, question ordering, questionnaire length, pretesting of the survey, and survey implementation. It is crucial that all tasks are executed properly for the survey to be a success.

## Data Compilation and Analysis

Once the sampling plan and survey have been implemented, the next step includes the compilation and analysis of the data. The analysis of the data may include the descriptive statistics of survey responses (e.g., the frequency distribution of survey responses) or it may include an analysis of the relationships between survey variables (e.g., determining whether customers with certain characteristics would be more or less likely to adopt a conservation measure). It is also possible that the survey data can be matched with the respondents' water billing histories. The combination of survey results with the water billing data can provide substantial information about how much water is used by various end uses.

# Estimate Potential Water Savings

The screening tests of applicability, technical feasibility, and social acceptability should reduce the number of measures that are to be fully evaluated. The next step in the screening process is to assess the potential water savings that might be expected from implementing a conservation measure. Estimates of water savings can be generated through either mechanical methods or through empirical methods. These are described in the following sections.

## Mechanical Estimates

Mechanical estimates have the advantage of being inexpensive and easy to obtain. This method, also known as the *engineering approach,* uses laboratory estimates or published data on water savings per installed device (or adoption of a given conservation practice). These data are combined with assumptions regarding the magnitude of factors expected to impact on the results of the conservation programs in order to develop estimates of program impacts. Therefore, mechanical estimates should not be considered an empirical evaluation technique. Rather, they are the best summation of a priori knowledge.

However, estimates of conservation savings obtained using the mechanical approach are quite sensitive to the underlying assumptions and relationships. The validity of these assumptions can easily come under attack, since they often rely on subjective conclusions and a great deal of professional judgment on the part of the analyst. Such assumptions are simply no substitute for actual field measurements, although through experience and careful analysis they may produce a fair approximation of reality.

Mechanical estimates may be considered appropriate for providing preliminary estimates of potential conservation savings when field measurements are not available. When field measurements are available, mechanical estimates may still be used to augment the statistical approach. Statistical models, on their own, may not be able to provide all of the impact estimates necessary for an evaluation. Furthermore, the engineering approach may be able to provide estimates for program participants that cannot be assessed by the statistical method. However, sole reliance on mechanical estimates for major investment decisions concerning the allocation of water resources should be discouraged.

The potential water savings resulting from a conservation measure can be estimated as a function of (1) the fraction reduction in water use, (2) the market penetration (or coverage), and (3) baseline water use. Because water conservation measures typically address a particular user sector (e.g., single-family residential, industrial establishments) and a specific water use dimension (e.g., indoor, outdoor, or peak use), the estimation of potential water savings is disaggregated on this basis. The water savings from a conservation measure (or the effectiveness of the measure) can be calculated as

$$E_{ijdt} = R_{ijd} * C_{ijt} * Q_{jdt}$$

where   $E_{ijdt}$ = effectiveness, or expected reduction in water use resulting from the implementation of measure $i$ (e.g., low-flow showerheads) in use sector $j$ (single-family sector) and water use dimension $d$ (winter/summer use or indoor/outdoor use) at time $t$ (e.g., 1990) in million gallons per day or acre-feet per year

$R_{ijd}$ = fraction reduction in water use from sector $j$ and dimension $d$ expected as the result of implementing measure $i$

$C_{ijt}$ = coverage of measure $i$ in sector $j$ and time $t$

$Q_{jdt}$ = unrestricted water use in sector $j$ and dimension $d$ and time $t$ in quantity per unit time (million gallons per day or acre-feet per year)

For example, water savings from a plumbing retrofit program in the single-family sector during winter in million gallons per day can be calculated based on the following values:

- A fraction reduction of 0.04 indicating that each retrofitted house saves 4 percent of water typically used during winter.
- Expected market penetration of 0.60 indicating that 60 percent of all single-family houses will install the retrofit devices.
- Baseline (or unrestricted) winter water use in single-family sector without the program of 50 million gallons per day.

Therefore, the estimated water savings from implementing the plumbing retrofit program would be 1.2 MGD [(0.04) ∗ (0.60) ∗ (50)]. Each of the parameters for estimating the potential water savings of conservation measures is discussed in the following sections.

**Unit Water Savings (Reduction).** The reduction factor measures the percent reduction in water use for a given user sector and dimension that is expected to result from a given conservation measure. For many conservation measures, empirical data on percent savings resulting from its implementation are unknown. The percent reduction can often be estimated based upon the calculation of unit water savings using device ratings and other engineering parameters. Given unit water savings in gallons per capita day (gpcd) or gallons per household per day (gphd), the following formula can be used to calculate the reduction factor

$$R_{id} = (S_{id}/Q_d)$$

where   $R_{id}$ = fraction reduction in water use from dimension $d$ expected as the result of implementing measure $i$
   $S_{id}$ = water savings in gallons per household per day in water use dimension $d$ (e.g., winter/summer water use or indoor/outdoor water use) resulting from measure $i$ (e.g., showerhead retrofit)
   $Q_d$ = average household water use in dimension $d$ without implementing the conservation measure in gallons per household per day

Table 9-5 presents generally accepted engineering parameters for selected conservation measures. Table 9-6 provides related information (e.g., toilet-flushing frequency, shower durations) for the estimation of water savings using engineering parameters. Conservation literature usually reports water savings per flush, per minute, or per load. In order to translate such estimates into total savings in million gallons per day, it is necessary to express the device savings as a fraction reduction. It should be noted that actual savings rates can vary considerably from the engineering estimates. Several factors can influence the actual savings of, for

example, a low-flow showerhead program. The obvious variables include
(1) water pressure at the plumbing outlets, (2) the characteristics of the former showerhead, (3) the degree to which shower valves are fully open
during showering, and (4) the degree to which consumers change their
habitual use of the fixture after it has been retrofitted. Somewhat less obvious variables that could influence water savings achieved in various communities are the demographic characteristics of the residential sector of
water users, such as average household size and family composition as
well as some socioeconomic variables such as income and education. It is
for this reason that empirical studies are necessary for deriving reliable
estimates of water savings for given study areas. However, when considering a conservation measure for potential implementation, it is necessary
for screening purposes to obtain preliminary approximations of the potential water savings. When no empirical data exist, it may be necessary to
rely on engineering estimates or the results of empirical studies conducted
by other agencies (see Table 9-7).

Table 9-8 provides estimates of the potential water savings of selected
conservation programs known as *Best Management Practices* (BMPs).
(These BMPs were developed for a Memorandum of Understanding
among water agencies in California coordinated by the California Urban
Water Conservation Council and are intended to provide a statewide stan-

**Table 9-5.** Engineering Parameters for Estimating Unit Water
Savings for Selected Conservation Measures

| Appliance/Fixture | Typical flow rates | Water savings |
|---|---|---|
| Standard toilet | 5.5 gal/flush | — |
| Low-flush toilet | 3.5 gal/flush | 2 gal/flush[+] |
| Ultra-low-flush toilet | 1.6 gal/flush | 3.9 gal/flush[+] |
| Toilet dam | — | 1 gal/flush |
| Toilet tank bag | — | 0.7 gal/flush |
| Toilet tank replacement bottles (2) | — | 0.5 gal/flush |
| Standard showerhead | 3.4 gal/minute | — |
| Low-flow showerhead | 1.9 gal/minute* | 1.5 gal/minute[+] |
| Standard washing machine | 55 gal/load | — |
| Water-efficient washing machine | 42 gal/load | 13 gal/load[+] |
| Standard dishwasher | 14 gal/load | — |
| Water-efficient dishwasher | 8.5 gal/load | 5.5 gal/load[+] |

SOURCES: Maddaus (1987), Brown and Caldwell (1984).

Note: See related parameters on Table 9-6.

*This is the rate after typical throttling. The current low-flow showerhead permitted
under California law is 2.7 gpm.

[+] Unit savings are relative to standard appliance and/or fixture.

**Table 9-6.** Engineering Parameters for Estimating Residential
End Uses of Water

| Parameter definition | Units | Likely range of average values |
|---|---|---|
| Indoor Uses | | |
| Average household size | Persons | 2.0–3.0 |
| Frequency of toilet flushing | Flushes/person/day | 4.0–6.0 |
| Flushing volumes | Gallons/flush | 1.6–8.0 |
| Fraction of leaking toilets | Percent | 0–30 |
| Showering frequency | Showers/person/day | 0–1.0 |
| Duration of average shower | Minutes | 5–15 |
| Shower flow rates | Gallons/minute | 1.5–5.0 |
| Bathing frequency | Baths/person/day | 0–0.2 |
| Volume of water | Gallons/bath | 30–50 |
| Washing machine use | Loads/person/day | 0.2–0.5 |
| Volume of water | Gallons/cycle | 45–50 |
| Dishwasher use | Loads/person/day | 0.1–0.3 |
| Volume of water | Gallons/cycle | 10–15 |
| Kitchen faucet use | Minutes/person/day | 0.5–5.0 |
| Faucet flow rates | Gallons/minute | 2.0–3.0 |
| Bathroom faucet use | Minutes/person/day | 0.5–3.0 |
| Faucet flow rates | Gallons/minute | 2.0–3.0 |
| Outdoor Uses | | |
| Average lot size* | Square feet | 5000–8000 |
| Average house size* | Square feet | 1200–2500 |
| Landscape area* | Square feet | 4000–5000 |
| Fraction of lot size in turf* | Percent | 30–50 |
| Water application rates* | Feet/year | 1–5 |
| Percent of homes with pools | Percent | 10–25 |
| Pool evaporation losses | Feet/year | 3–7 |
| Frequency of refilling pools | Five years | 1–2 |
| Frequency of car washing | Times/month | 1–2 |

SOURCES: Brown and Caldwell (1984), Boland et al. (1990), Dziegielewski (1990a).
* Reflects single-family averages.

dard for the evaluation of urban water conservation programs.) The estimates of water savings present the best available estimates of what can be
achieved by the adoption of the BMPs.

**Market Penetration (Coverage).**   The aggregate water savings in a
given water use sector will depend on the proportion of the sector actually

**Table 9-7.** Example Calculation of Reduction Factor Using Engineering Parameters

The following calculations represent the water savings expected as the result of a showerhead retrofit program. The savings rate represents a difference in average winter water use between homes with low-flow showerheads and homes without low-flow showerheads.

Nonconserving showerhead flow rate = 3.4 gallons/minute
Low-flow showerhead flow rate = 1.9 gallons/minute
Estimated showering time = 4.8 minutes/person/day
Average winter household water use = 200 gallons per household per day
Average household size = 2.5 persons
Water use with nonconserving showerhead = 3.4 gal/min * 4.8 min/
person/day = 16.3 gpcd
Water use with low-flow showerhead = 1.9 gal/min * 4.8 min/person/
day = 9.1 gpcd
Water savings = 16.3 gpcd – 9.1 gpcd = 7.2 gpcd

At an average household size of 2.5 persons, the savings rate would be 18.0 gallons per household per day (2.5 persons * 7.2 GPCD). The formula for calculating the reduction factors representing the fraction of, for example, single-family winter water use is

$$R = (18.7 \text{ GPHD})/(200 \text{ GPHD during winter}) = 0.09 \text{ (or 9 percent)}$$

affected by the conservation measure. The market penetration (or coverage) is an indication of the percent of water use within a given sector affected by a given measure at a given point in time. Coverage factors may be approximated by the percent of water users who are in compliance with, or who have adopted, a given measure.

The coverage factor is an unknown quantity for most measures. Typically, the estimation of coverage is based upon the expected compliance rates (e.g., plumbing codes) or rates of installations (e.g., voluntary ultra-low-flush toilet installations). For example, a ULF toilet retrofit program is expected to achieve 10,000 installations. The 10,000 installations can be converted into market penetration by recognizing that on average each single-family home has 1.9 toilets. If there are 60,000 single-family homes in the service area, then the market penetration (or coverage) is about 9 percent:

$$(10,000) \div (60,000 * 1.9) = 0.088$$

For screening purposes, surveys of the general public (see previous discussion on social acceptability) may provide indications of the potential initial market penetration. Alternately, market penetration can be deemed as a

**Table 9-8.** Potential Water Savings for Selected Conservation Practices in California

| Measure description | Estimated water savings | | |
|---|---|---|---|
| | GPCD | GPHD | Percent |
| Single-family retrofit (pre-1980 homes) | | | |
|    Toilet retrofit | 1.3 | 4 | 1% of annual use |
|    Low-flow showerhead | 7.2 | 22 | 4% of annual use |
| Multifamily retrofit (pre-1980 homes) | | | |
|    Toilet retrofit | 1.3 | 3 | 1% of annual use |
|    Low-flow showerhead | 7.2 | 17 | 6% of annual use |
| Home water audits (pre-1980 single-family homes) | | | |
|    Toilet retrofit | 1.3 | 4 | 1% of annual use |
|    Low-flow showerhead | 7.2 | 22 | 5% of annual use |
|    Leak repair | 0.5 | 2 | <1% of annual use |
|    Outdoor use | — | — | 5–10% of outdoor use* |
| Home water audits (post-1980 single-family homes) | | | |
|    Low-flow showerhead | 2.9 | 9 | 2% of annual use |
|    Leak repair | 0.5 | 2 | <1% of annual use |
|    Outdoor use | — | — | 5–10% of outdoor use* |
| Large landscape water audits | — | — | 10–20% of irrigation use in affected sectors |
| Landscape requirements for new commercial, industrial, multifamily complexes | — | — | 10–20% of irrigation use in affected sectors |
| Governmental plumbing retrofit | — | — | 5% of indoor use in governmental sectors |
| Distribution system water audits and leak detection | — | — | <10% of total production |
| 1992 California plumbing code | | | |
|   Residential use (relative to pre-1980 housing units) | | | |
|     Toilets | 16 | 50 | 10% of annual use |
|     Showerheads | 7.2 | 22 | 4% of annual use |
|   Nonresidential use (relative to pre-1980 housing units) | | | |
|     Toilets | — | — | 3% of annual use |
|     Public facilities | — | — | <1% of annual use |

SOURCES: California Urban Water Conservation Council (1993); Brown and Caldwell Consultants (1991); and Planning and Management Consultants (1991).

Notes:
(1) gpcd = gallons per capita per day; gphd = gallons per household per day.
(2) Assumes 3.1 persons per household in single-family units and 1.9 persons per housing unit in multifamily units.
(3) California Plumbing Code became effective in 1978 and was revised in 1980; therefore water savings will vary between pre- and post-1980 homes.
(4) *Provides best available estimates, but not substantiated with empirical data.

policy goal for a given conservation measure (e.g., a goal to achieve a target of 80 percent of single-family units for a retrofit program). Consideration must be given to the difference between consumers who are exposed to a conservation measure (or who receive conservation devices) and those consumers who actually implement the conservation measure. The value of the coverage factor may be expected to either increase over time as more users comply with a given measure or decrease as the water-saving devices wear out with time or as customers remove the devices (e.g., toilet dams).

**Unrestricted Water Use.** The product of the reduction and coverage factors ($R_{ijd} * C_{ijt}$) of a given conservation measure will provide an estimate of the percent savings from the implementation of a conservation measure. In order to estimate aggregate sector water savings over time, it is necessary to apply this estimate of percent savings to unrestricted water use forecasts (i.e., unrestricted by potential water demand management programs) for a given sector for each given point in time. It should be noted that the product of reduction and coverage may vary over time due to changing compliance rates and consumer behavior and the attrition of devices. Therefore, in order to evaluate the effectiveness of water conservation measures in future years, it is necessary to prepare an unrestricted water use forecast by user sector (see discussion in Chapter 4).

## Empirical Estimation

Although mechanical methods are generally used to develop initial estimates of water savings from potential water conservation programs, these estimates can be refined by conducting empirical evaluations of pilot program results. The purpose of an empirical estimation will be to obtain accurate measurements of changes in water use that are attributable to the conservation program. The selection of appropriate data and methods for the impact evaluation will depend on specific objectives. It is generally accepted that the simpler methods of impact evaluation are the least costly, yet the most unreliable. The more data-intensive and sophisticated methods of impact evaluation are more costly (in time and effort), but their results tend to be more reliable. Therefore, the desired reliability of estimates will influence the selection of data and methods of analysis. (For more specific detail on the subject of program evaluation methods, see Chapter 10, Demand Management Program Evaluation Methods.)

## Candidate Measures

Once the conservation measures have been tested for applicability, technical feasibility, social acceptability, and potential water savings, a list of

candidate conservation practices, devices, and regulations can be prepared. The candidate measures can thereby be included as components of
water conservation programs. Such a list should also include information
on expected water savings and the degree to which each practice or device
is expected to be acceptable to water users. The estimates of potential
water savings of conservation measures can be rank-ordered with the
least likely candidates excluded from further evaluation.

## Define Implementation Conditions

The previous screening tests of applicability, technical feasibility, social
acceptability, and potential water savings should have reduced the number
of measures that are to be fully evaluated. In order to provide a basis for
further evaluation of each conservation practice, it is necessary to formulate each measure as a fully developed *conservation alternative or program*. It
is helpful to distinguish between a conservation measure (method, practice, or technique) and a conservation program or alternative. A conservation measure is broadly defined to include any activity, practice,
technological device, law, or policy that can potentially reduce water use.
The definition of a conservation measure should be sufficiently narrow to
permit the evaluation of its applicability, technical feasibility, and social
acceptability. However, a conservation program or alternative is designed
to facilitate implementation of one or more conservation measures.

Therefore, the following sections discuss the types of implementation
conditions that must be specified before the conservation alternatives can
continue to be evaluated. The implementation plan of each program
should include the following elements:

1. Program contents
2. Definition of the target population and program participants
3. Program incentives
4. Customer contact mode(s)
5. Schedule of program implementation and duration
6. Specification of responsible agencies
7. Program evaluation plan

### Program Contents

The contents of the potential water conservation alternatives must be
clearly defined. Some programs may be designed to accomplish the
adoption of only one conservation measure, while others may be

designed to promote several conservation measures. For example, an Ultra-Low-Flush (ULF) Toilet Replacement Program listed in Table 9-9 may cover only the installation of the ULF toilet. Other programs may include or promote packages of individual conservation devices and practices. For example, a High Use Home Water Audit Program may include several conservation devices, educational literature, oral instructions, and even simple plumbing services. Packaging of conservation measures will often enhance the cost-effectiveness of a conservation program.

The following is an example of a package of devices and activities covered by a home water audit program (Brown and Caldwell 1990):

**Table 9-9.** Examples of Water Conservation Programs

Type of program/Program name

Educational Programs
    Elementary School Conservation Education Program
    High School Conservation Education Program
    Mass-Media Public Information Program
    Xeriscaping Garden Demonstration Program

Plumbing Retrofit Campaigns
    Residential Plumbing Fixture Retrofit Program
    Governmental Plumbing Retrofit Program
    ULF Toilet Replacement Program

Conservation Audits
    High Use Home Water Audit Program
    Commercial and Industrial Audit Program
    Residential Leak Detection Program
    Distribution System Audit and Leak Detection Program
    Large Landscape Irrigation Audit Program

Conservation Ordinances
    Indoor Plumbing Code for New Construction
    Point-of-Sale Plumbing Fixture Ordinances
    Residential Landscaping Ordinance
    Commercial Landscaping Ordinance
    Submetering Ordinance
    Water Waste Ordinance

Landscape Replacement
    Buy-Back Turf Program
    Xeriscape Replacement Program
    Irrigation Retrofit Program

SOURCE: Dziegielewski et al. (1993).

Indoor Audit

- Measure indoor water flow rates
- Checks for leaks
- Survey frequency of appliance/fixture use
- Install low-flow showerheads, toilet dams, and faucet aerators

Outdoor Audit

- Obtain soil probe core
- Check root development
- Determine moisture distribution
- Test sprinkler flow rate
- Determine turf type
- Determine soil absorption rate
- Make recommendations on watering times and frequencies based on preceding observations and calculated evapotranspiration rates

The potential costs of a conservation program will be affected by the advertising and/or public information component of the program. Therefore, in devising the program content, consideration must be given to program publicity. Various channels of communication can be used for program publicity. The channel of communication to be used for a particular program should be chosen with regard to the target population. For example, if a citywide water-rationing program is being implemented, the program announcements should be made through the mass media (i.e., television, radio, newspaper). Alternately, a plumbing retrofit program targeting only selected neighborhoods could mail notification letters only to the customers in targeted neighborhoods.

## Definition of Target Population and Program Participants

The program design should clearly establish the target population for the program. For example, a home water audit program can be directed to:

1. All residential dwellings within the service area
2. All residential single-family homes built prior to the implementation of a plumbing code
3. All homes occupied by households with annual income not exceeding a certain amount
4. All single-family homes with annual water use exceeding 120,000 gallons

In addition to type and age of housing, household income, and water usage, the target population can be defined in terms of water use type, geographic location, meter size, or other characteristics.

Once the program contents and target population have been determined, it is necessary to determine the coverage of the conservation program (e.g., 2000 single-family homes in a particular neighborhood; 25,000 multifamily units citywide; or top 20 percent of total industrial water use). When estimating the coverage of the program, the planner must consider that not all eligible consumers in the target population will participate in the program. For example, only 50 percent of eligible single-family residences contacted for a home water audit may actually participate. Although the entire target population may be initially contacted, thereby affecting notification costs, the actual number of participants will affect the field labor and equipment costs. As mentioned previously, the coverage of a conservation program can be deemed as a policy goal for a given conservation program (e.g., to achieve the retrofit of 80 percent of all single-family homes).

## Program Incentives

Savings in water use alone may not be a sufficient incentive for program participation. Other incentives are often added in order to increase the number of willing and eligible program participants. Typically, economic incentives for motivating consumers to conserve water include (1) rebates, (2) tax credits, or (3) subsidies. If the conservation program being screened uses economic incentives, the amount and conditions of the incentives should be specified in the tentative implementation plan. For example, ULF toilet programs typically utilize rebates to encourage the purchase and installation of the toilets. The maximum dollar amount of the rebate should be set at the difference between the price of a typical conserving toilet and the price of a typical ULF toilet, plus a reasonable margin.

## Customer Contact Modes

Almost all conservation programs will require some level of contact with each eligible customer of the target population. Such contacts can be facilitated in several ways, including:

1. Telephone solicitation and scheduling
2. Call-in requests and scheduling
3. Sign-up booths in malls or at public events

4. Door-to-door canvassing

5. Direct written contact

6. Mass-media contacts

These contact modes (or distribution modes) differ in terms of cost and the potential rate of success in reaching the target population. The selection of a specific mode of customer contact will depend on the desired goals of program participation and the type of conservation program.

## Schedule of Program Implementation and Duration

A realistic schedule for implementing all phases of the program should be developed. This should include both the estimated start date of the program as well as its duration. The duration of the program can be specified as a time frame (two-year program) or as an expected implementation rate (5000 single-family homes to be retrofitted each year for the next five years).

## Specification of Responsible Agencies

For each conservation program considered for implementation, there is likely to be various levels of involvement from different agencies, organizations, and individuals. Some conservation programs may be conducted with water agency personnel, and others may be contracted out to private companies. Some conservation programs may require the recruitment of field labor from temporary help agencies or community groups. The participation of various groups in the program implementation phase will affect program costs. Therefore, the specification of the responsible agencies in each conservation program must clearly be defined in the implementation plan.

## Program Evaluation Plan

If the conservation program being considered is expected to be the subject of an empirical program evaluation (see Chapter 10), this needs to be stated in the program implementation plan. The program evaluation plan should include a clear description of the programs' goals and data collection methods. The costs involved with conducting an empirical program evaluation may need to be considered in the benefit-cost analyses.

## Conduct Benefit-Cost Analysis

The previous sections draw attention to the need to water clarify conservation goals and to assess the acceptability, feasibility, applicability, and potential water savings of candidate measures. However, it is also necessary to conduct an evaluation of the economic merit of proposed water demand–management programs. That is, there is a point at which efforts to conserve water resources require more resources than are saved. In a world of increasingly limited budgets, it becomes increasingly important to choose the most efficient programs from that set of programs that are economically viable. Benefit-cost analysis provides a screening mechanism for choosing the most efficient management alternatives. The concept of using economic analysis for analyzing the potential for water demand–management alternatives was initially posed by Baumann et al. (1980). Carefully formulated conservation alternatives can be compared in terms of their cost and their yield in savings. Furthermore, benefit-cost analysis enables supply augmentation alternatives (e.g., the construction of a new reservoir) to be compared with demand management alternatives (e.g., conservation) using the same economic criteria. A benefit-cost analysis also enables planners to evaluate the effect of water conservation efforts on the sizing and timing of future water facilities.

Benefit-cost analysis and cost-effectiveness are two terms that are sometimes used synonymously. However, there is a distinction between the two. In benefit-cost analysis both the costs and benefits are examined. Cost-effectiveness refers only to a comparison of the costs of various actions, the benefits being assumed equal. However, benefits of projects are rarely equal especially with water demand management alternatives. Benefit-cost analysis allows comparison of alternatives of differing levels of benefits.

Benefit-cost analysis is conceptually very simple, but its application can become quite involved because of the many benefit and cost elements that must be accounted for. In this section, some of the benefit and costs that are typical of conservation program analysis will be described. Examples of the costs of some retrofit programs will be presented. A discussion of the different analytical perspectives from which benefit-cost estimates can be calculated will be introduced.

### Demand Management Costs

Each water conservation alternative is associated with a proposed implementation plan. The implementation plan indicates the agency or agencies responsible for implementation of the conservation program, the time of implementation, the coverage of the measure, and the specific actions that must be taken to implement and maintain the measure. These details are used to estimate implementation costs, including those occurring at the time of implementation and those required to maintain the program's full effectiveness in the future. Implementation costs are also estimated separately by the type of organization to which they accrue. Costs borne by a water utility (i.e., utility program costs) are stated separately from those borne by water users (i.e., customer program costs), by public interest groups, or by public agencies other than the water utility. Each water conservation alternative must also be reviewed to determine whether any additional costs can be expected to result from its implementation. When such effects can be found, their future levels should be estimated and evaluated. Typical conservation program costs can be categorized as follows:

*Utility Program Costs:* The process of implementing a conservation program requires expenditures on such items as labor, materials, advertising, and economic incentives paid to program participants.

- Administration
- Unit costs (retrofit kits, etc.)
- Field labor
- Incentives
- Publicity
- Program evaluation

*Decreased Utility Revenue:* Without rate adjustments, reduced water use leads to reduced revenues (particularly in short-term emergency drought conditions). It should be noted that in most cases, long-term conservation measures have small year-to-year impacts on revenues and that other revenue requirements tend to drive compensating rate increases. However, decreased utility revenue can be a cost to the water supply utility.

*Customer Program Costs:* The customer participating in the program may be required to bear the cost of materials, installation, and operations and maintenance costs.

- Equipment, materials, installation
- Operation and maintenance

*External Costs:* External costs are those that are not typically included in the cost accounting methods and can be difficult, if not impossible, to quantify. For example, it may be important to consider the effects of a conservation program on aesthetic values (e.g., decreased customer satisfaction due to the replacement of lush green lawns with xeriscaping) or upon the environment.

**Example 1: An Ultra-Low-Flush Toilet Retrofit Program.** A water utility implemented an ultra-low-flush toilet distribution program. The cumulative program costs for a four-month period are presented in Table 9-10. The same data are presented graphically in Figure 9-1. Note that these figures do not include the management cost internal to the water utility. However, that cost is quite small. Note especially, the declining total cumulative cost of distributing the toilets. After the first month, 4285 toilets had been distributed at a per unit cost of $206.85. By the end of the fourth month, 20,423 toilets had been distributed at an average cost of $136.49 per toilet (the incremental unit cost of the 20,424th toilet was $117.16). The breakdown of the total cost provides an indication of the factors contributing to the decreasing unit cost. The major contributors to the overall decrease for the four-month period are the decreasing costs of program design, marketing, management, and warehousing. Other costs have remained nearly constant.

**Example 2: A Showerhead Retrofit and Conservation Education Program.** A water utility is planning a showerhead retrofit and conservation education program to be conducted through elementary schools within its service area. With respect to the education program, a contractor to the water utility will develop conservation education curricula for grades K to 3 and for grades 4 to 6. The contractor will work closely with curriculum directors and with teachers. The students will be provided with games to play and a survey to complete with the help of their parents; both the games

**Table 9-10.** Ultra-Low-Flush Toilet Retrofit Program—Costs

| Program activity | Month 1 | Month 2 | Month 3 | Month 4 |
|---|---|---|---|---|
| Cumulative ULF toilets distributed | 4,285 | 9,550 | 13,929 | 20,423 |
| Cumulative cost per ULF toilet: | | | | |
| 1. Program design, development, marketing and management support | $85.26 | $45.57 | $37.64 | $30.60 |
| 2. Payment to community-based distribution organization | 20.63 | 19.01 | 20.97 | 21.49 |
| 3. Payment for recycling old toilet | 7.94 | 6.19 | 7.13 | 6.00 |
| 4. Warehousing cost for ULF toilet inventory | 27.58 | 16.22 | 13.01 | 10.70 |
| 5. Purchase of toilet and related materials | 65.44 | 64.91 | 66.79 | 67.70 |
| Total cumulative unit cost of distributed ULF toilets | $206.85 | $151.90 | $145.54 | $136.49 |

SOURCE: Planning and Management Consultants, Ltd. (1994).

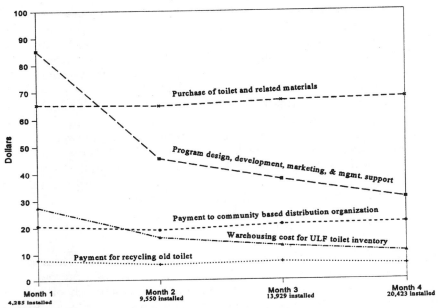

**Figure 9-1.** Ultra-Low-Flush Toilet: Retrofit Program—Actual Costs

and the survey will be designed to promote an understanding of water use and to encourage conservation behavior. If requested by the parents, the students will be given retrofit kits to take home. Each kit will contain one or two low-flow showerheads, one or two toilet tummies, leak detection tablets, and literature designed to encourage installation. The goal is to contact 300,000 students in the hope that 100,000 kits will be installed.

The projected costs of the retrofit program are presented in Table 9-11 for three different levels of distribution. The most noticeable feature of this cost breakdown is the considerable reduction in cost per kit that can be realized with a wider distribution. The cost per kit projected for a distribution of 100,000 kits is less than one-half of the cost projected for a distribution of 33,300 kits. Note also that this reduction arises because of the administrative costs that are projected to be constant (in total) for any level of distribution. The costs of the showerheads, the toilet tummies, the adapters, and the dye tablets are relatively constant on a per kit basis.

It is important to note that the administrative cost estimates in Table 9-11 are not typical of a showerhead retrofit program. This program has been designed for an area where there is already a relatively high saturation rate. It is costly to contact the remaining noninstallers and to convince them to install the low-flow showerheads. Much of the administrative (and "other") costs arise from the education and promotion efforts.

**Table 9-11.** Showerhead Retrofit Program—Cost Estimates

| | | | |
|---|---|---|---|
| Students contacted | 100,000 | 200,000 | 300,000 |
| Kits delivered | 33,300 | 66,600 | 100,000 |
| Showerhead | $106,590 | $213,180 | $323,000 |
| Toilet insert | 42,075 | 84,150 | 128,000 |
| Adapters | 5,049 | 10,098 | 15,000 |
| Dye tablets | 5,000 | 10,000 | 15,000 |
| Restocking charge | 20,100 | 10,050 | |
| Administration costs | 577,487 | 577,487 | 577,487 |
| Other costs | 111,663 | 145,706 | 181,575 |
| Total cost | $867,964 | $1,050,671 | $1,240,062 |
| Cost per kit | $26.06 | $15.78 | $12.40 |

SOURCE: Planning and Management Consultants, Ltd. (1994).

Note: Other costs include printing, warehousing, travel, and communication.

## Demand Management Benefits

The benefits of water conservation measures arise from the reduction in water use and/or water losses. Some of the benefits are related to the reduction in water supply costs. These types of benefits require knowledge about existing water supply costs as well as planned future expansions in water supplies and/or facilities (i.e., a water supply plan). Other benefits are indirectly related to reductions in water use and do not require knowledge about the water supply plan (e.g., energy cost savings).

Implementing water conservation alternatives may reduce the short-run incremental costs of water supply and wastewater disposal. Short-run incremental costs are those costs that are immediately changed in response to changing use patterns, and are not associated with capital facilities. They include the costs of chemicals, energy, labor, materials, and the like. These costs should be identified separately for existing and planned facilities, including wastewater collection, treatment, and disposal facilities as well as water supply facilities.

Long-run incremental costs of water supply and wastewater disposal may also be reduced. Therefore, the water supply plan must also be analyzed to determine the expected long-run incremental cost associated with the last increments of capacity that are to be provided throughout the planning period. Long-run incremental costs are those costs associated with providing capital facilities for water supply and wastewater disposal. These costs vary with the design capacity of the facilities; design capacity varies as a consequence of changes in patterns and levels of water use. The foregone costs of new supply facilities may be a major benefit of the conservation program.

Typical conservation program benefits can be categorized as follows:

*Utility Cost Savings:* In the short run, a conservation program will reduce the cost to the utility for chemicals, energy, labor, and water purchases. In the long run, a conservation program can lower the cost of capital facilities for water supply and wastewater disposal.

- Reduced purchases of water
- Reduced operation and maintenance costs
- Deferred, eliminated, or downsized new facilities

*Program Participant Benefits:* In the short run, a conservation program will reduce the cost to the utility for chemicals, energy, labor, and water purchases. In the long run, a conservation program can lower the cost of capital facilities for water supply and wastewater disposal.

- Incentives
- Reduced water bills
- Reduced wastewater bills
- Reduced energy bills

*System Reliability:* Water supply systems cannot always meet the demands placed on them. For example, in periods of drought, some systems may not be able to deliver all the water people normally require. Such shortages can be quite costly; shortages can disrupt industrial production and landscaping efforts. To the extent that conservation programs are effective, the reliability of the system is enhanced.

*External Benefits:* Some benefits lie outside the standard accounting methods. For example, water conservation reduces extraction of water from rivers and groundwater. Increased supplies can have significant benefits for fish, wildlife, and water quality. These benefits can be difficult, if not impossible, to quantify.

Benefits of water conservation programs accrue separately to water, energy, and wastewater utilities and to the utilities' customers. Careful accounting can allocate the benefits to the proper beneficiary. Not all benefits can be quantified, and nonquantifiable benefits must be dealt with outside of the quantitative economic analysis.

## The Accounting Perspective

As presented earlier, there are a variety of benefits and costs that are typically of interest to the water utility planner. However, an added complication arises when aggregating all of the benefits and costs. That is, a benefit from one point of view may be a cost from another point of view. This issue is referred to as the *accounting stance* or the *accounting perspective*.

As an example of differing accounting perspectives, consider an incentive paid by a utility to promote the installation of ultra-low-flush toilets. If the benefit-cost analysis is conducted from the viewpoint of the utility, then the incentive payout is a cost; if conducted from the perspective of the program participant, then the payout is a benefit. Thus, at the start of the analysis, it is necessary to be clear about the perspective to be taken. Five different accounting perspectives are usually defined, although most benefit-cost studies rarely consider all five (O'Grady 1993).

*Participant:* Benefits and costs are defined from the point of view of the program participant. All residents of the utility's service area may not participate in any given program.

*Utility:* The public or private water utility can be expected to have a different perspective on benefits and costs than do the program participants.

*Community:* This perspective takes the point of view of the local community which includes both the utility and the conservation program participant. However, this is not a simple summation of utility and participant costs and benefits. Nonparticipating ratepayers are also part of the community perspective.

*Ratepayer:* In some cases a conservation program may affect water rates. This effect can be captured in the ratepayer's perspective.

*Society:* The societal perspective considers benefits and costs that include, but also go beyond, the community perspective. State or national interests may be included.

For any conservation program it is useful to consider several perspectives; this provides a broader picture of the acceptability of the proposed program. This is important because any benefit-cost analysis may face scrutiny from several interests such as the utility, voters, lending agencies, and city planners. A benefit-cost analysis that has considered all these perspectives may be much more defensible. At a minimum, a well-prepared analysis will provide an indication as to who will support or not support a particular conservation program. For example, a voluntary retrofit program may be seen as highly beneficial from the utility's perspective. However, if the targeted participants perceive no benefit, it is unlikely that the program will succeed. Table 9-12 provides a summary of benefits and costs by account perspective.

## Accounting for Intangibles

Any conservation program may yield benefits and costs that are very hard to quantify in terms of dollars and cents. For example, reduced lawn watering may result in lawns that are less green and therefore aestheti-

**Table 9-12.** A Summary of Benefits and Costs by Accounting Perspectives

| | Participant |
|---|---|
| Benefits | + Program Participant Benefits, including incentives received from any source<br>+ System Reliability |
| Costs | − Customer Program Costs |

| | Ratepayer |
|---|---|
| Benefits | + Utility Cost Savings<br>+ System Reliability |
| Costs | − Utility Program Costs, including incentives paid by utility |

| | Utility |
|---|---|
| Benefits | + Utility Cost Savings |
| Costs | − Utility Program Costs, including incentives paid by utility<br>− Decreased Utility Revenues |

| | Community |
|---|---|
| Benefits | + Utility Cost Savings<br>+ System Reliability<br>+ Program Participant Benefits, including all incentives received |
| Costs | − Utility Program Costs, including incentives paid by utility<br>− Customer Program Costs |

| | Society |
|---|---|
| Benefits | + Utility Cost Savings<br>+ Program Participant Benefits, excluding all incentives<br>+ External Benefits<br>+ System Reliability |
| Costs | − Utility Program Costs, excluding all incentives<br>− Customer Program Costs<br>− External Costs |

SOURCE: Planning and Management Consultants, Ltd. (1994).

cally less pleasing; a city's effort to conserve water may gain it political goodwill from state or national conservation agencies; reduced water consumption may increase in-stream flows that may benefit the recreation industry or the environment; delivery system reliability may be enhanced as a result of reduced water consumption. Although there have been improvements in the techniques for evaluating some of these benefits and costs, they are still very difficult to quantify.

Even though such costs or benefits as mentioned previously are difficult to quantify, they may be very important considerations in the decision-making process. However, unless these items can be quantified satisfactorily, it is best to leave them out of the benefit-cost analysis. In this manner, the benefit-cost analysis can be scrutinized, and its soundness judged, on the basis of its treatment of tangible benefits and costs. This does not mean that intangibles should be ignored in the decision to proceed with a conservation program. Indeed, they may be major elements in the decision to undertake a conservation program. As a separate part of the decision-making process the results of the benefit-cost analysis can be weighed in the balance with the more subjective and intangible elements of the process. To attempt to include intangibles in the benefit-cost analysis will simply obscure the overall decision.

## The Process of Discounting

The benefits and costs of conservation programs typically accrue over an extended period of time. As such, it is necessary to discount the value of future benefits and costs to make them comparable. Discounting future costs and benefits is an important part of the benefit-cost analysis. It allows the comparison of options with very different patterns of costs and benefits over time. If all else is equal, everyone would prefer to have their benefits earlier and costs later. Discounting to present worth is the method that enables recognition of the *time preference.*

Underlying discounting is the concept that a benefit experienced in the present is worth more than if it were to be experienced in the future. For example, given a choice between receiving $100 today or receiving $100 several years from now, most people would choose to receive the money today (even if waiting carried no risk).

The degree to which a present option is preferred over a future option can be characterized by a discount rate. Figure 9-2 graphically depicts how the present value of $100 declines the further in the future it is received (assuming either a 4 percent or 8 percent discount rate). It is important to note that the notion of discounting should be kept separate from the effects of inflation. In the example immediately preceding, the preferred choice would remain the same with or without inflation. In standard benefit-cost analysis, inflation usually is ignored; this is because it is assumed that the inflation rate affects benefits and costs equally. Usually, this assumption is fairly realistic (however, if you have information that the inflation rate for benefits is not the same as for costs, then it is important to account for inflation).

The choice of a rate for discounting the value of future benefits and costs is an important one, particularly when the project involves a large initial capital expenditure, or when the magnitude of benefits and costs is expected to change over the planning period. The discount rate should

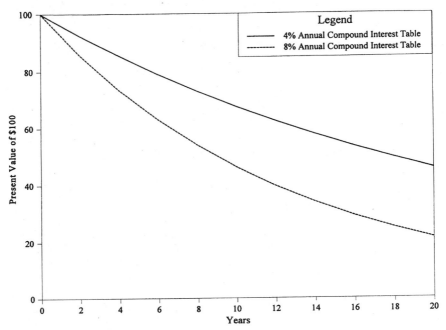

**Figure 9-2.** Present Value of $100 Received in the Future

reflect the true alternative use of the capital expenditures, or the *opportunity cost*. One school of thought holds that the discount rate should equal the cost of capital in the private sector. However, this requires the assumption that the alternative use of all funds used for the project would be invested in the private sector if the project were canceled. The ultimate sources of funding for the projects considered by many water utilities are taxes and water rates. It should not be assumed that every extra dollar collected for taxes and water sales would otherwise flow into capital investment; a good deal of it would be spent by individuals on current consumption. An annual rate of discount that acknowledges this reallocation would be lower than the cost of capital and may fall between 0 and 6 percent.

## Benefit-Cost Measures

One of the primary goals of a water conservation program is to be economically efficient. The larger the savings and smaller the costs to implement, the more favorable the program. However, it is worth noting that in some cases, the costs of a conservation program may be too great even to be considered for implementation (due to budgetary constraints), regardless of the potential benefits. If this is the case, those programs that are outside budgetary limitations can be eliminated from further analysis.

Alternately, in some cases (i.e., water supply shortages), conservation programs must be implemented, although it is not beneficial in an economic sense (e.g., benefit-cost ratio not greater than one). In this case, benefit-cost analyses can be used to rank the most economically optimal and effective program. This section describes two methods of evaluating water conservation programs using the notions of costs and benefits developed in the prior sections. These methods take into account the time value of money.

Two of the most commonly used methods of economic analyses performed to evaluate relative costs and benefits are (1) net present value and (2) benefit-to-cost ratio. These two methods follow the guidelines of the Standard Practice Manual for the *Economic Analysis of Demand-Side Management Programs* of the California Energy Commission (1987). While each analysis is correct and uses similar input, the output is expressed differently. All methods are valid analyses, but some agencies may have a preference as to which method is used. Having collected the background economic data (costs and benefits), it is possible to evaluate water conservation alternatives using both economic analysis methods. These methods can be used to evaluate one particular conservation measure or a number of measures being implemented together.

**Net Present Value.**   The net present value (NPV) method provides a comparison of costs and benefits throughout the life of the conservation project. The NPV of a project is calculated as the difference between the net present value of benefits and the net present value of costs. A project with a positive (>0) NPV is economically viable. The following equation is used to perform this analysis.

$$\text{Net present value} = \sum_{t=1}^{n} (B_t - C_t)/(1 + i)^t$$

where   $B_t$ = all applicable benefits in year $t$
$C_t$ = all applicable costs in year $t$
$i$ = selected interest (discount) rate
$n$ = number of years in the time period selected for analysis

**Benefit-to-Cost Ratio.**   The benefit-to-cost ratio is another commonly used method of measuring economic efficiency. This method determines the ratio of the net present value of benefits to the net present value of costs. Those conservation measures with a benefit-to-cost ratio greater than 1.0 are economically efficient. The method of calculation is

$$\text{Benefit-cost ratio} = \sum_{t=1}^{n} \frac{B_t}{(1+i)^t} \bigg/ \sum_{t=1}^{n} \frac{C_t}{(1+i)^t}$$

# Integrate Water Conservation into Water Supply Plans

Evaluation of water conservation alternatives results in a list of conservation programs with all advantageous and disadvantageous effects identified and measured or described for each measure. In order to integrate water conservation programs into water supply plans, individual programs must be combined to form water conservation proposals (or demand management plans); the proposals become the water conservation elements of the supply plans. Water conservation proposals can be developed in order to enhance desired features of the final water supply/conservation plan. The following sections describe the development of water conservation proposals suitable for integration into a water supply plan, illustrated by Figure 9-3.

## Proposal Development Principles*

**Merit Order.**   Because of the possibility of interactions among individual water conservation measures and/or programs, it is helpful to introduce individual programs into each alternative water conservation proposal in merit order—the *best* program is included first, followed by the *next best,* and so forth. The definition of *merit* depends upon the objective of the water conservation proposal. For example, a proposal intended to maximize net benefits implies a different notion of merit than does a proposal directed to other objectives, such as reducing wastewater outflows. Various objectives of the water conservation proposal might include the following:

1. *Economic Objective.* For purposes of developing a water conservation proposal that makes the maximum net contribution to the economic objective, water conservation programs can be arranged in order of decreasing net benefits. Net benefit is defined as the sum of all advantageous economic effects less the sum of all disadvantageous effects.

---

* These principles were initially developed by Baumann et al. (1980).

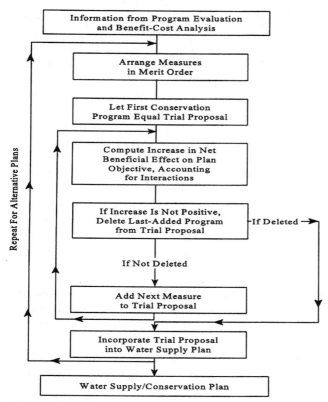

INTEGRATION OF WATER CONSERVATION
INTO WATER SUPPLY PLANS

**Figure 9-3.** Integration of Water Conservation into Water
Supply Plans (Dziegielewski et al. 1993)

2. *Environmental Objective.* For purposes of developing a water conservation proposal that makes the maximum contribution to the environmental objective of water conservation, programs can be arranged in order
   of decreasing net environmental benefit. Net environmental benefit is
   defined as the sum of all advantageous environmental effects less the
   sum of all disadvantageous effects. If environmental effects are diverse,
   considerable judgment may be required to achieve this ranking.

3. *Combined Objectives.* Other plans may be proposed that affect significant
   trade-offs between the economic and the environmental objectives.
   Such plans are judged according to a selected combination of the two
   objectives. Water conservation programs are arranged in decreasing

order of their individual contributions to the same combination of objectives.

**Interaction Effects.** Combinations of water conservation measures and programs can be expected to exhibit interactions with respect to both water conservation savings and implementation costs. In some cases, interactions may also appear for other advantageous and disadvantageous effects, including environmental effects.

Interactions with respect to effectiveness (or water savings) can result when two different conservation measures impact the same water use or water use behavior. In fact, whenever metering and pricing measures are implemented in conjunction with other water conservation measures, interactions can be expected.

Interactions with respect to implementation costs can result when two measures share common implementation characteristics. Typically, the implementation of two measures at the same time results in costs borne by the water utility and/or public agencies that are less than the sum of costs of implementing the measures individually. In most cases, joint implementation can be expected to reduce aggregate implementation costs. This interaction is most striking in the case of educational efforts.

**Net Beneficial Effects.** As individual water conservation measures are added to trial water conservation proposals, the net beneficial effect of adding the additional measure must be determined. In every case, the net beneficial effect is defined with respect to the plan objective—economic, environmental, or combined. The net beneficial effect can be found by determining the excess of all advantageous effects of the plan objective over all disadvantageous effects of the plan objective before adding the additional measure, and then determining the same excess after adding the additional measure. Finally, it may be noted whether the second amount is greater (an increase in net beneficial effect) or less (a decrease in net beneficial effect) than the first.

## Development of Alternative Conservation Proposals

**Economic Objective.** The proposal that makes the maximum net contribution to the economic objective is developed from the list of eligible water conservation programs, arranged in suitable merit order and evaluated on the basis of the water supply plan. Proposal development begins by choosing the first listed program. Then, the second program is added to the first. The advantageous effects of the second program are added to those of the first along with the disadvantageous effects. Interactions between the two plans are investigated, and the summed effects adjusted when necessary. For example, if the two programs interact with respect to water savings such that their combined effectiveness is less than the sum of their effectiveness, advantageous economic effects must also be adjusted downward.

If the water conservation proposal now formed (two programs) exhibits a net contribution to the economic objective (net beneficial effect) that is larger than that recorded for the immediately preceding plan (one program), the second program is retained and the development proceeds. If the proposal development proceeds, additional programs are tentatively added in the same way, effects are summed, interactions are investigated, summed effects are adjusted when necessary, the net beneficial effect is tested. Development stops when the next program in the merit order list fails to contribute to the net economic effect: this is when the contribution to the economic objective is maximized. The programs then included constitute the water conservation element of the water supply/conservation plan that maximizes economic benefits.

**Environmental Objective.** The proposal that makes the maximum contribution to the environmental objective is developed from the list of eligible water conservation measures, arranged in suitable merit order, and evaluated on the basis of the water supply plan. The second program is then added to the first. Tentatively, the advantageous effects of the second program are added to those of the first, and the disadvantageous effects are added. Interactions between the two programs are investigated, and the summed effects adjusted when necessary.

If the water conservation proposal now formed (two measures) exhibits net beneficial effects on the environmental objective that are judged to be not less than those observed before the first measure was added, the second measure is retained and the development proceeds. If not, the second measure is removed and the development stops. If proposal development proceeds, additional measures are tentatively added in the same way; effects are summed, interactions are investigated, summed effects are adjusted when necessary, and the results are tested. Development stops when the next measure in the merit ordered list reduces the cumulative net beneficial effect on environmental objective. The resulting conservation proposal constitute the water conservation element of the water supply/conservation plan that maximizes environmental benefits.

**Combined Objectives.** Combined objectives reflect trade-offs between the planning objectives or between one of the planning objectives and other considerations. Conservation programs can be merit-ordered, and the conservation proposal developed in a manner analogous to that described earlier for the basic plans. Effects of individual programs are estimated on the basis of the other water supply plan. The merit order should reflect the combined objectives. For example, if a compromise between economic and environmental objectives is sought, measures with net beneficial effects on both objectives would be listed first, followed by those considered less desirable in view of both objectives, and so forth. The proposal formulation then proceeds until the mix of environmental and economic effects desired in the plan cannot be enhanced by adding other measures.

**Treatment of Potentially Feasible or Potentially Acceptable Measures.**    Some eligible water conservation measures may not be feasible or acceptable under existing physical or social conditions. These measures are categorized as *potentially feasible* or *potentially acceptable,* and the conditions under which they would become feasible in the future are specified. Initially, such measures may be included in the list of eligible measures, and in the development of water conservation proposals. Whenever one of the final water supply/conservation plans includes potentially feasible or potentially acceptable measures, however, a second plan will be developed on the same criteria, except that potentially feasible and potentially acceptable measures will be excluded from the list of eligible measures. Both plans will be presented for comparison, so that the consequences of not implementing the potentially feasible or potentially acceptable measures can be contrasted to the difficulty of removing impediments.

A good example of treating potentially feasible and acceptable measures is the selection process of Best Management Practices by urban water districts in California (as per the signed Memorandum of Understanding Regarding Urban Water Conservation in California; California Urban Water Conservation Council 1993). Two types of practices are distinguished: *present* and *potential.* The present best management practices are conservation methods for which water savings, economic, environmental, and social effects are being documented in field applications. Documented savings from these practices will be incorporated into the overall water supply planning program of the agencies participating in the program. The potential best management practices are those with uncertain outcomes that require the development of technical, economic, and social acceptability data before a major commitment of resources for their implementation can be made.

## Supply Reliability Considerations

The advantages of water conservation result largely from possible reductions in supply capability, when system reliability is held constant. If the overall reliability of the supply system is altered by the implementation of water conservation practices, additional disadvantageous or advantageous effects are created. The need to identify and measure these additional effects can be avoided by holding system reliability constant throughout the analysis. Following development of alternative water conservation proposals, this assumption should be tested by determining the performance of each alternative supply plan, with and without the water conservation element, for the last year in the planning period assuming design drought conditions. Supply plans with water conservation will differ from those without this element in having downsized or delayed construction schedules as well as lower levels of water use. When water deficits appear under design drought conditions, emergency water use reduction measures (not already incorporated in the water conservation proposals) are required. The extent and severity of measures required for supply plans that incorporate conservation should not exceed those for the corresponding supply plans without conservation.

## Documentation of Water Management Plans

The procedures described in the previous sections will result in one or more water conservation proposals that can be integrated with water supply plans to form water supply/conservation plans. Whenever proposals include potentially feasible or potentially acceptable programs, alternative plans will be developed that exclude these measures. The documentation of each water supply/conservation plan must include:

1. A full list of water conservation programs considered, showing conservation measures that were excluded from these programs as not technically feasible, those that were excluded as not socially acceptable, those that were excluded as not eligible because of negative impacts on the economic and environmental objectives, and those that were excluded in the process of plan formulation.

2. A list of water conservation programs considered not applicable because they are already implemented, or because definite commitments have been made to implement them within the planning area.

3. A list of each water conservation program included in the proposal, with a full description for each program, an indication of the agency or other entity responsible for its implementation, and a summary of the implementation plan including estimated coverage and duration.

4. Aggregate implementation cost for the water conservation proposal, expressed as annualized cost; implementation cost for the proposal identified by responsible party (utility, residential water users, etc.).

5. Aggregate water savings for the water conservation proposal, shown separately with respect to average-day water use, maximum-day water use, and average-day sewer contribution; shown for selected times throughout the planning period.

6. A description of the water supply plan, without water conservation, including a summary of beneficial and adverse effects.

7. A description of the water supply/conservation plan, incorporating the water conservation proposal, including a summary of beneficial and adverse effects.

8. A summary of the performance of the water supply plan (without conservation) and the water supply/conservation plan for the last year of the planning period under design drought conditions. Data provided should include projected supply capability, projected water use (including maximum-day use), and the nature and assumed effectiveness of emergency water use reductions measures required, if any.

# References

Barakat and Chamberlin, Inc. 1994. *Marin Municipal Water District Water Efficiency and Conservation Master Plan.* Corte Madera, CA.

Barakat and Chamberlin, Inc. 1994. *Conservation Measure Technology Profiles: Portland Area Water Providers.* Portland, OR.

Barakat and Chamberlin, Inc. 1995. *Conservation Program Descriptions: Portland Metropolitan Area Water Providers.* Portland, OR.

Baumann, Duane D., John J. Boland, and John H. Sims. 1980. *The Evaluation of Water Conservation for Municipal and Industrial Water Supply: Procedures Manual.* Ft. Belvoir, VA: U.S. Army Corps of Engineers, Institute for Water Resources.

Boland, John J., Alexander A. McPhail, and Eva M. Opitz. 1990. *Water Demand of Detached Single-Family Residences: Empirical Studies for the Metropolitan Water District of Southern California.* Carbondale, IL: Planning and Management Consultants, Ltd.

Brown and Caldwell Consulting Engineers. 1984. *Residential Water Conservation Projects Summary Report.* Pleasant Hill, CA: Department of Housing and Urban Development.

Brown and Caldwell Consulting Engineers. 1991. *Assessment of Water Savings from Best Management Practices.* Los Angeles, CA: Metropolitan Water District of Southern California.

Brown and Caldwell Consulting Engineers. 1990. *Assessment of Water Conservation Potential in Metropolitan's Service Area.* Pleasant Hill, CA.

California Energy Commission, Energy Efficiency and Local Assistance Division and California Public Utilities Commission, Division of Ratepayer Advocates. 1987. *Economic Analysis of Demand-Side Management Programs: Standard Practice Manual—Staff Report.* Sacramento, CA.

California Urban Water Conservation Council. 1993. *Memorandum of Understanding Regarding Urban Water Conservation in California.* Sacramento, CA.

Dziegielewski, Benedykt, Eva Opitz, and Dan Rodrigo. 1990a. *Seasonal Components of Urban Water Use in Southern California.* Los Angeles, CA: Metropolitan Water District of Southern California.

Dziegielewski, B., E. Opitz, J. Kiefer, D. Baumann, M. Winer, W. Illingworth, W. Maddaus, and P. Macy. 1993. *Evaluating Urban Water Conservation Programs: A Procedures Manual.* Denver, CO: American Water Works Association.

Frey, J. H. 1988. *Survey Research by Telephone.* Newbury Park, CA: Sage Publications.

Fowler, Floyd J. 1989. *Survey Research Methods.* Newbury Park, CA: Sage Publications.

Maddaus, William O. 1987. *Water Conservation.* Denver, CO: American Water Works Association.

O'Grady, K. L. 1993. "Methods for Analyzing Benefits and Costs of Conservation Alternatives." Paper prepared for Conserv93, a conference sponsored by the American Water Works Association, Las Vegas, December 1993.

Planning and Management Consultants, Ltd. 1991. *Municipal and Industrial Water Use in the Metropolitan Water District Service Area: Interim Report No. 4.* Los Angeles, CA: Metropolitan Water District of Southern California.

Planning and Management Consultants, Ltd. 1994. *Evaluating Urban Water Conservation Programs: Workbook.* Los Angeles, CA: Metropolitan Water District of Southern California.

Seattle Water Department. 1992. *Water Supply Plan.* Seattle, WA.

# 10

# Demand Management Program Evaluation Methods

**Benedykt Dziegielewski**
*Southern Illinois University
at Carbondale*

**Eva M. Opitz**
*Planning and Management
Consultants, Ltd.*

Today, most major urban water suppliers fund water conservation activities. Many conservation initiatives are undertaken in order to prevent waste and to encourage customers to use water efficiently. Economic evaluation of demand-side alternatives is needed for justifying public and private expenditures on water conservation. In places where such programs operate on small budgets and are treated by the management as public relations programs there is no perceived need to evaluate their effectiveness. In times of financial difficulties such programs are often discontinued. They may be revived when a rate increase causes a flood of telephone calls from dissatisfied customers.

## The Role of Program Evaluation

When conservation programs are undertaken as part of long-term water management plans, the need for their evaluation is as important as the

**283**

need to plan for a new reservoir or a new pipeline. Their costs and performance must be known before major investments in water conservation are undertaken. Certainty of water savings must be established to ensure the inclusion of demand-side programs among other alternatives of integrated resource planning. No water utility would spend $50 million on a new reservoir if the amount of supply that could be obtained was unknown. The cost of evaluating water conservation programs is no different from the cost of planning and designing new supply projects. Decision makers are very reluctant to spend money on a large-scale conservation program with an uncertain outcome.

Before going into a detailed discussion of the needs and objectives of program evaluation, it is important to answer the question: What is program evaluation? Because a program is a coordinated group of things to be done, its evaluation is a form of audit that determines whether the program implementation plan was followed and what the outcome was. Accordingly, program evaluation can be defined as an inquiry that attempts to answer a series of questions about the program's performance. The success or failure of a program can be measured in many ways. For example, a residential, on-site water audit program may fail to produce water savings in a cost-effective manner but it may succeed in establishing good public relations with residential customers and it may win public support for other conservation programs. Therefore, the questions to be answered by program evaluation must consider the many aspects of the program pertaining to its impacts on water use, costs, customer response, logistics, and other outcomes.

The most critical questions in program evaluation are:

- Has the program reduced water use?

- How much water use reduction was achieved?

- What was the actual cost of the program?

- Has the change in water use persisted over time?

- How many customers have participated in the program?

- Were the implementation procedures adequate?

- What are the reasons for nonparticipation and/or dropping out of the program?

The cost of evaluating water conservation programs may be very high and must be justified in terms of the payoff. An adequate evaluation will allow conservation planners to: (1) formulate least-cost conservation programs for achieving various levels of water savings; (2) implement only those programs that are cost-effective (i.e., programs with long-term benefits exceeding costs); and (3) take into account conservation effects (i.e., reduction in water demand) in planning for water supply development. In many instances, the overall payoff from expenditures on program evaluation will be to reduce the cost of water supply and increase its reliability in the long run. Improvements in water use efficiency along with well-timed investments in water supply should result in lower water rates paid by the consumers as compared to rates without the benefits of water conservation.

The level of uncertainty surrounding program outcomes must be known in order to allow water planners to fully incorporate conservation alternatives into their long-term water supply plans. A careful examination of empirical evidence of achievable water savings, public acceptability of water conservation measures, and other conservation impacts will increase the level of confidence among planners and managers who are responsible for the provision of an adequate water supply to urban economies.

The objective evaluation of the program's performance, which consists in the rational (or scientific) measurement of program outcomes is not the only reason for undertaking program evaluation. It is important to recognize the promotive, financial, administrative, or political motives behind program evaluation. In cases where the motives are other than rational, it may be difficult to maintain scientific objectivity in performing the evaluation. However, evaluations that fail to achieve the standards of objectivity may have far-reaching adverse consequences on any future conservation efforts by undermining public confidence in an agency's decision-making abilities.

## Objectives of Program Evaluation

The questions of program evaluation can be recast in the form of specific objectives. Usually the most important objective is to determine whether the program achieved its intended goals. This objective would include reasons for exceeding the goals or for failing to achieve them. In some cases it may be necessary to modify the program in order to improve the chances of its success. Typically, the purpose of program evaluation would be:

1. To compare program effects against intended goals
2. To determine whether an ongoing program needs modification
3. To identify the effects of the program on long-term needs for resources and facilities
4. To document program implementation and performance for designing and evaluating future programs

## Process Evaluation versus Impact Evaluation

The purpose of *process evaluation* is to track the operational efficiency of the program. This is done in order to obtain measurements of the effectiveness of program implementation methods and to assess the overall effects of the program. The process evaluation design should cover all elements and effects of the conservation program, except its impacts on water use. In some cases (especially where funds are limited), the results of the process evaluation may be combined with engineering estimates of water savings (or with results of previous evaluation studies) to generate estimates of aggregate savings from the conservation program. Typically, interim (or formative) evaluations are used in process evaluations. Table 10-1 shows a list of potential measurements in process evaluation. Customer surveys are the main source of information on market penetration. Table 10-2 gives examples of potential survey topics.

**Table 10-1.** Potential Measurements for Process Evaluation

| Questions | Potential measurements |
| --- | --- |
| Who? | Total number of eligible customers (i.e., target group) |
| | Characteristics of eligible customers (e.g., single-family homes built prior to 1980, commercial establishments with large turf areas) |
| | Number of eligible customers receiving notification (percent) |
| | Number of eligible customers contacted during fieldwork (percent) |
| | Number of contacted customers who agreed to participate in the program |
| | Characteristics of participants (e.g., family size, age, income, type of establishment) |
| How? | Detailed procedures of conducting program (e.g., delivery modes, incentives, advertising) |
| What? | Number of contacts with program participants (and outcomes of each) |
| | Number of devices (of each type) installed |
| | Number of and type of services performed |
| | Number of audits/devices delivered per field worker (per day or per week) |
| | Customer satisfaction |
| Where/when? | Location of participants (geographically) |
| | Economic environment (e.g., water and wastewater prices) |
| | When program took place (day, month, year) |
| How much? | Costs of providing program (e.g., equipment/devices, labor, transportation, supplies) |

## **Table 10-2.** Potential Survey Topics

(A) Preprogram Survey
  (1) Belief about the effectiveness of potential conservation program
  (2) Beliefs about the economic benefits of potential conservation program
  (3) Preferred implementation methods of potential conservation program
      (e.g., retrofit program: depot, mail, door-to-door distribution)
  (4) Beliefs about the fairness of potential conservation program

(B) Postprogram Survey
  (1) Whether conservation program service or device was received
  (2) Whether conservation program service was implemented or device
      installed (what components? how many?)
  (3) Whether conservation program recommendations or devices were main-
      tained
  (4) Satisfaction with program services or devices
  (5) Satisfaction with program implementation methods

(C) Pre- and Postprogram Surveys
  (1) Socioeconomic (household/business characteristics)

  *If residential:*
      (a) Type of residence (single-family, duplex, townhouse, etc.)
      (b) Ownership of residence
      (c) Family size and age of family members
      (d) Home value (or contract rent)
      (e) Socioeconomic status (income, education, occupation)
      (f) Lot size and type of yard (lawn, garden, etc.)

  *If business:*
      (a) Type of business (by Standard Industrial Classification)
      (b) Primary service or product
      (c) Number of employees (number of hotel rooms, number of students)
      (d) Size of operation (production or business volume)

  (2) Information on water-using fixtures and appliances
  (3) Information on water-using activities and frequencies
  (4) Awareness of conservation activities
  (5) Attitudes about water conservation
  (6) Other conservation practices

The most critical component of evaluation is the measurement of savings in water use. The purpose of *impact evaluation* is to obtain accurate measurements of changes in water use which are clearly attributable to the conservation program. Additional objectives may include attributing water savings to each element (or conservation measure) of the conservation program. The decisions to be made based on the results of impact evaluation are much more important than those associated with the evaluation of the implementation process. Therefore, greater precision of measurements is very desirable. Because the data required by impact evaluation may need special collection and/or processing, the cost of the impact analysis may be very high. Impact evaluation procedures are likely to be final.

# Data Collection for Process Evaluation

The need for early planning cannot be overstated with respect to the process evaluation. The sources of information for the process evaluation must be determined well in advance of program implementation. In most cases, selected data parameters for the process evaluation are gathered during the delivery of services. Therefore, data collection forms and procedures must be prepared prior to program implementation, and program implementation staff must be trained in the data collection methods. The process evaluation can be conducted by the program implementation staff or by an independent reviewer (e.g., contract consultant). It should be noted that independent evaluations of conservation programs will provide added validity to findings (especially when the conservation programs are issues of public and political debate). Assuming that the program will be evaluated by an independent reviewer, a series of interviews with program implementation staff should be conducted. The interviews should involve staff responsible for various aspects of program implementation, including project management, customer contact, and data tracking. The purpose of the interviews is to develop full understanding of the implementation process. A review of any contracting arrangements for equipment or services should also be documented.

## Data Collection for Impact Evaluation

Various types of data can be used in making the comparison of water use and other characteristics within the study-design framework. Water utilities usually have one or more production meters that are read at least daily. Because there are generally only one or two production meters, they are regularly maintained for accuracy, and therefore usually produce highly reliable measurements of water flows into the distribution system. A record of total water sales during each billing period can also be used to estimate conservation savings if (1) the sales record is of sufficient quality, and (2) the sales are broken down by homogeneous customer groups (e.g., single-family residential, multiunit residential, commercial, industrial, public, and other classes of users). The usefulness of billing history of individual customers depends on the frequency and continuity of customer meter readings. The best billing history data are obtained by reading meters once a month and not using estimated consumption in place of actual readings. Databases can be built from billing records by supplementing them with data from other sources, such as random samples of customers. Laboratory performance data of water-efficient hardware can be used to calculate differences in water use rates between various water-using appliances. However, the application of laboratory data in estimating the effects of conservation programs often poses many problems.

## Sample Selection for Evaluation

Water utilities have the ability to monitor the water use of all customers, classes of customers or individual connections. In statistical terms, studies involving all users would represent the use of the entire population. However, the study of an entire population must limit the number of measurements on each customer due to the cost constraints and may not be capable of producing answers to a number of research questions. Because of the cost and other considerations, our knowledge is almost invariably based on samples or fragments of total population.

When selecting samples for evaluation it is important to consider several questions that are important for obtaining valid results from the survey. These questions include:

- What do we want to learn from the sample?
- What group of customers is to be represented by the sample?
- What characteristics of customers should be measured?
- What amount of error can be tolerated?
- What sample size is needed?
- What listings of customers should be used?
- How should the sample be selected?

Scientific sampling designs specify methods for sample selection and estimation of sample statistics which provide, at the lowest possible cost, estimates that are precise enough for the study objectives. Probability sampling refers to any sampling procedure that relies on random selection and is amenable to the application of sampling theory to validate the measurements obtained through sampling. Popular sampling plans for selecting a probability sample include simple random sampling and stratified sampling. Simple random sampling involves selecting $n$ sampling units out of a population of size $N$, such that every one of the distinct samples (where each sample consists of $n$ sampling units) has an equal chance of being drawn. In stratified sampling, the sampled population of $N$ units is first divided into two or more subpopulations called strata. If a simple random sample is taken from each stratum, then the sampling procedure is described as stratified random sampling. A stratified sample can produce greater precision of the estimates for the same sample size drawn from a heterogeneous population.

The size of the sample depends on the precision of measurement that is required and the variance in the parameters to be estimated. Although efficient sample sizes can be calculated using statistical relationships between variance, precision, and sample size, in many studies sample sizes are selected using researchers' judgment. Sample sizes, based on judgment, range from 500 to 2000 residential customers and 50 to 500 nonresidential customers.

## Design of Program Evaluation Studies

Once a conservation practice is adopted by a program participant, the baseline demand which represents water use without the practice cannot be directly measured, and the unaltered demand has to be somewhat reconstructed. In practice, all study designs employ comparisons of water use behavior (and other customer characteristics in some cases) over time and/or between groups of customers. Possible types of comparisons are illustrated in Figure 10-1. The time continuum is divided into two periods by the implementation of the conservation program, the behavior and characteristics of water users before the program implementation (or pretreatment) and after program implementation (or posttreatment). The treatment, or implementation of the conservation program, also divides the water use into two groups: the control group of nonparticipant and the treatment group of program participants. As a result of program implementation, four categories of conditions are created:

1. Pretreatment conditions of control group

2. Posttreatment conditions of control group

3. Pretreatment conditions of treatment group

4. Posttreatment conditions of treatment group

Several comparisons can be made among these categories. For example, a sample of water users drawn from the treatment group after the program

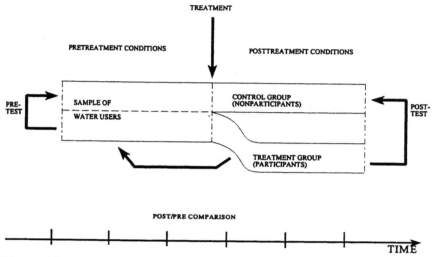

**Figure 10-1.** A Basic Experimental Design

can be compared to itself before the program. The treatment group after the program can also be compared to a sample representing the control group after the program. A number of other comparisons can also be made.

## Validity of Evaluation Studies

Determining design validity is the most important consideration in developing an evaluation procedure. An appropriate evaluation design (i.e., experimental design) enhances the validity of results and provides the analyst with maximum information within the constraints of the budget of the evaluation project. If the evaluation design is flawed, it may be difficult or impossible to salvage any meaningful results from the data.

A valid experimental design will provide definitive answers to the relevant research questions. The ability to generalize the findings of a particular evaluation to other programs and consumer groups depends on whether the evaluation design is capable of isolating and controlling: (1) characteristics of the program being evaluated, (2) characteristics of the targeted customer groups, and (3) external influences of weather, drought, and/or price changes.

The evaluation process can be affected by a number of factors. Although some factors are beyond the control of the analyst, in most cases the obscuring factors are inherent to the research method employed. Factors that can threaten the internal and external validity of evaluation designs include:

1. *History* or changes in the environment surrounding the program

2. *Maturation* caused by change in participant attitude over time

3. *Strategic bias* arising when program participants behave differently when they know they are being examined

4. *Instrumentation* or the effect of changing the calibration of the measuring instrument (e.g., "nonblind" data collectors)

5. *Regression-to-the-mean* caused by choosing the participants on the basis of extreme measurements

6. *Self-selection bias* when program participation is voluntary

7. *Experimental mortality* when program participants drop out of the program

Although these factors are design-specific (e.g., history and maturation do not apply to designs which use cross-sectional data), they have the potential of biasing the results of the program evaluation unless some appropriate actions (to quantify their effects) are taken by the analyst.

## Measuring Program Impact

The critical questions of impact evaluation are:

- How to measure water savings?
- How to project savings into the future?
- What is the level of uncertainty surrounding the estimates of savings?
- Are the projected water savings reliable enough to justify the cost of a demand management program and to be included in long-term water supply plans?

In assessing the reliability of the estimates of water savings, it is helpful to separate the problems of measurement of savings from the problems of projecting savings into the future. With respect to measurement, one must be able to measure the actual change in water use that results from a water demand management program in question. In projecting water savings 20 or 30 years into the future one must determine whether the observed change in water use will persist over time. Measurement deals with actual data on water use and other variables which influence water use. Water utilities already have water use records and have to identify and quantify the changes in water use that can be clearly attributed to the water demand management program. In projecting future savings one must deal with projected water use and assess the impacts of a program and other influencing factors on future water use. The reliability of the measurements of current water savings must be known in order to infer the reliability of the estimates of savings in the future. Because the future is inherently uncertain, the reliability of the estimates of projected conservation savings will always be lower than the reliability associated with the measurement of existing savings.

A precise measurement of water savings is difficult because the observed water use often shows great variability among users and it also significantly varies over time for the same user. For example, the amount of water used inside and outside a residential home can vary substantially from month to month and from household to household. This variability is caused by many factors including changes in weather, season, household size, composition, income, prices of water and wastewater, development of water leaks, and other factors. Therefore, the most important consideration in measuring water conservation savings is the design of a measurement procedure that is capable of correctly measuring not only the changes in water use but also separating these changes into those caused by the program and those caused by changes in weather, prices, economic factors, and other factors.

Reliability projections of future effects of demand management programs cannot be made without first determining future water use without such programs. The rates of water use, such as gross per capita use, are a function of many factors. These rates may decrease, increase, or remain unchanged depending on future values of these factors and their effects on water use. Therefore, the effects of the demand management programs sponsored by the area water agencies must be measured from the future baseline demands which incorporate the effects of these other factors.

All these influences make the baseline water use a moving target. In order to determine the reliability of the projected savings of the utility-sponsored conservation programs, one must first determine the reliability of water-demand forecasts. If it is assumed that per capita use rates will remain constant over time, the forecasts will not likely produce reliable estimates. Furthermore, if the uncertainty surrounding the estimates of future water use is unknown, then the reliability of future water savings also remains unknown.

## Data and Methods for Measuring Program Impacts

The selection of appropriate data and methods for program evaluation depends on the specific measurements that are needed. For example, time-series regression analyses of aggregate data can be used to measure gross savings of large-scale programs or the combined conservation effect of all conservation activities in the service area. If estimates of water savings attributable to specific conservation measures are desired, then customer-level data (i.e., billing histories) are most appropriate. The desired reliability of estimates will also influence the selection of data and methods of analysis. Generally, analyses of aggregate data will provide estimates that are less reliable than those obtained from analyses of customer-level data.

Engineering methods and statistical comparisons are inexpensive and simple to utilize. However, these techniques are based on stringent assumptions that are often violated. The multivariate regression approach is shown to be more appealing because of the wealth of information the method uses to explain variance in household water use. Finally, the combination of engineering and statistical methods allows researchers to verify and validate the measurements of water conservation savings.

## Using Engineering Estimates

Engineering (or mechanical) estimates are obtained using laboratory measurements or published data on water savings per device or conservation practice. These data can be combined with assumptions regarding the magnitude of factors expected to affect the results of the demand management program in order to generate estimates of program savings. However, the resultant estimates can be very sensitive to the assumptions regarding the frequency and intensity of fixture use. In case of toilet use, depending on the assumptions about flushing volumes and frequency of flushing, the resultant savings can range from 19.5 gallons per person per day (3.9 gallons × 5 flushes per person per day) to 7.6 gallons per person per day (1.9 gallons × 4 flushes per person per day). The high estimate is almost 3 times greater than the low estimate. The validity of the assumptions used in the preceding example can easily come under attack, since they rely on subjective conclusions and a great deal of professional judgment of the engineer or analyst.

Despite their obvious shortcomings, engineering estimates may be considered appropriate for providing preliminary estimates of potential savings when field measurements are not available. They can also be used to verify statistical estimates by setting limits on a possible range of savings. However, they become most useful in leveraged techniques where they can be used to augment and strengthen statistical models.

## Using Statistical Comparisons

Statistical comparison methods produce estimates of savings by comparing water use between a participant group and a control group (or by comparing water use before and after the program). The comparison of means method is derived from the statistical theory of randomized controlled experiments that utilize a treatment/control design. Conservation savings are estimated as the difference in the mean level of water use between the treatment group and the control group.

When adhering to a strict experimental design, the comparison of means method is more likely to produce meaningful and reliable results in situations where (1) the expected conservation effect is large when compared to mean water use, (2) the variance in water use is small, (3) the mean and variance in water use are very similar (in terms of size) for both groups prior to treatment, and (4) the sample sizes in the treatment and control groups are large. In summary, the comparison of means method can produce reliable and informative results if used in conjunction with experimental designs and large sample sizes.

An analysis of water savings attributable to a pilot residential retrofit program that conformed to the treatment/control evaluation design was undertaken for the city of Tampa, Florida (Kiefer and Davis 1991). The results of this comparison are shown Table 10-3. All water use accounts that remained active between January 1989 and December 1990 were matched with mail survey data from a sample of treatment and control households. Average annual water use per household was derived from a sample set of 1363 households. Out of this group, average annual water use was calculated for 1085 treatment households and 278 control households. The table shows annual average household water use for each group in gallons per day (GPD) before and after the retrofit program; standard deviations are in parentheses.

Mean water use in the treatment group was below that of the control group in both pre- and posttest periods. Using a statistical procedure, estimates of mean water use in the treatment and control groups were compared for both periods in order to deduce whether these differences were attributable to chance or if the two samples came from populations of unequal means. A statistical test indicated that the two means were essen-

**Table 10-3.** Example of Statistical Comparison: Comparison of Annual Average Household Water Use (in Gallons per Day)

|  | Preretrofit 1989 | Postretrofit 1990 | Difference (savings) |
|---|---|---|---|
| Treatment ($n = 1085$) | 228.4 (157.1) | 192.0 (130.4) | −36.4 |
| Control ($n = 278$) | 247.4 (182.8) | 212.9 (153.5) | −34.5 |
| Difference (savings) | −19.0 | −20.8 | −1.9 |

tially the same during the preretrofit period. In other words, the observed differences in the mean of the two samples were not large enough to conclude that water use in the two groups differed. However, a similar test of significance indicated that the means of the treatment and control groups during the postretrofit period were different. It was therefore concluded that even though average water use for both groups had declined over the two time periods under study, water use in the treatment group had declined *more* than water use in the control group, and that this difference was attributed to other than chance. The subsequent difference in water use of 1.9 GPD could be taken to be a measure of the reduction in water use due to the retrofit, given that the two groups reacted similarly to other external factors that occurred during the study periods.

## Using Multivariate Regression

Multivariate regression models represent the most sophisticated method of comparing water use data over time or between groups of customers while controlling for the effects of a large number of external factors. One can choose from a variety of regression methods depending on the types of available data and the acceptable level of estimation complexity.

A time-series observation of the volume of water sold in consecutive billing periods can be used to measure conservation effects of full-scale programs while controlling influences other than the program. The reliability of the estimates will depend on: (1) the ability to disaggregate sales data into classes of similar users (e.g., single-family residential, small commercial), (2) the ability to separate (or account for) the seasonal effects and weather effects in the time series data, and (3) the ability of the estimation technique to deal with nonconstant error variance and correlation of model errors through time. The effects of the program under investigation can be measured by including an indicator variable which separates the time series data into pre- and postprogram periods. However, it is also important to capture the effects of other passive and active conservation measures which are adopted by water customers independently of the program under evaluation.

If customer-level monthly (or billing period) data on water use for a period of two to four years can be supplemented with information on customer characteristics and such external factors as price of water and weather conditions, then a pooled time-series cross-sectional data set can be constructed and used to estimate the parameters of a multiple regression model. The important explanatory variables will depend on the customer class. In the residential sectors they usually include information on family characteristics, household features, and frequency of outdoor use. This measurement technique can produce very accurate estimates of actual water savings for some programs, especially those targeting the residential sector.

The major drawback of multivariate regression models is that they are relatively expensive and time-consuming. They require large amounts of data and greater expertise on the part of the analyst. They also need large sample sizes and are less appropriate for nonresidential water.

An example of a regression model is presented in Table 10-4. The model was used to estimate water conservation savings attributable to the indoor plumbing retrofit program conducted by the city of Phoenix Department of

**Table 10-4.** Example of Multiple Regression OLS Model:
PHOENIX RETROFIT PROGRAM

| Variable name | Mean value | Regression effect |
|---|---|---|
| Cooling degree days | 387.86 | 0.45 |
| Home area (square feet) | 2650.36 | 0.05 |
| Yard type[a] | 2.23 | 66.44 |
| Swimming pool[b] | 0.42 | 66.54 |
| Persons per household (number) | 3.23 | 25.49 |
| Home value ($1000) | 63.97 | 1.36 |
| Green area (square feet) | 3980.03 | 0.00 |
| Frequency of yard watering (per week) | 1.95 | 9.09 |
| Washing machine[c] | 0.98 | 83.38 |
| Frequency of dishwasher use[d] | 1.52 | 8.27 |
| Toilets per residence | 2.02 | 34.14 |
| Yard watering[e] | 0.88 | 39.71 |
| Ownership of residence[f] | 0.92 | 39.72 |
| Sewerage system[g] | 0.95 | −40.26 |
| Date of structure[h] | 0.83 | 23.04 |
| Retrofit installation[i] | 0.44 | −14.90 |
| Frequency of washing machine use[j] | 3.07 | 2.05 |
| Constant | — | −102.56 |

Sample size = 28,285 pooled observations.
$R^2 = 0.292$ $F = 684.911 (0.000)$.
Gallons per day = 496.637.
[a] Yard type (0 = no yard; 1 = desert; 2 = desert/green; 3 = green).
[b] Swimming pool (0 = no; 1 = yes).
[c] Washing machine (0 = no; 1 = yes).
[d] Frequency of dishwasher use (0 = zero/don't know; 1 = 1–2 loads; 2 = 3–4 loads; 3 = 5–6 loads; 4 = 7–8 loads; 5 = 9–10 loads; 6 = 11–5 loads; 7 = 16–20 loads; 8 = 21–25 loads; 9 = 26 or more loads).
[e] Yard watering (0 = no; 1 = yes).
[f] Ownership of residence (0 = rent; 1 = own).
[g] Sewerage system (0 = septic; 1 = sewer).
[h] Date of structure (0 = built since 1980; 1 = built prior to 1980).
[i] Retrofit installation (0 = no; 1 = yes).
[j] Frequency of washing machine use (0 = zero/don't know; 1 = 1–2 loads; 2 = 3–4 loads; 3 = 5–6 loads; 4 = 7–8 loads; 5 = 9–10 loads; 6 = 11–15 loads; 7 = 16–20 loads; 8 = 21–25 loads; 9 = 26 or more loads.

Water and Wastewater in the summer of 1985. Considering the "pooled" nature of the data set, the goodness-of-fit ($R^2$) coefficient of 0.292 indicates a reasonably good fit of the model. The estimated regression coefficient for the installation variable is −14.90, indicating that households that installed retrofit devices during the campaign use, on average, 14.90 gallons per day less water than households without such devices. The 95 percent confidence interval for this regression coefficient with a standard error of 3.9 GPD is from 7.2 GPD to 22.6 GPD. The Phoenix model provided satisfactory estimates of conservation savings that were confirmed by several independent analyses.

### Combined Evaluation Methods

Leveraged approaches combine the statistical and engineering methods. The accumulated experience with program evaluation indicates that it is very difficult to obtain measurements of water savings with a high level of precision using a single best method. The most precise estimates can be achieved by taking advantage of the strong features of the three previously discussed approaches by using alternative data sets and estimation techniques to estimate the effects of the same program. Although the results of each approach may differ, the known strengths, weaknesses, and biases of each approach can be used to narrow down the confidence bands surrounding the actual water savings.

The most promising method of leveraging information involves the use of information from one approach within the procedures of another. For example, engineering estimates or special metering measurements can be used as independent variables in statistical models. Unfortunately, there are no documented cases of the use of this approach in the water industry.

## Uncertainty in Measuring Water Savings

Statistical comparisons and multivariate regression methods produce information on the uncertainty surrounding the estimated values of water savings. For example, 95 percent confidence intervals can be constructed using the standard error of the estimated difference between mean water use in treatment and control groups and the critical value of the $t$-statistic of 1.96 (for two-sided test of significance at the 0.05 level of probability). For example, if the mean and standard deviation of water use in the treatment group are, respectively, 409.5 and 343.1 gallons per day (GPD) and in the control group 438.7 and 372.3 GPD, the difference between means is

−29.2 GPD. With sample sizes of 1200 for the treatment group and 800 for the control group the standard error of this difference can be computed— 16.2 GPD. Using this information the upper and lower limits of the 95 percent confidence interval are calculated as: savings = (−29.2 ± 1.96 * 16.2) = −29.2 ± 31.8 GPD. This confidence interval extending from 2.6 to −61.0 is very wide. It is important to note here, that the precision of the estimates of water use (as opposed to water savings) is substantially greater. For example, in the treatment group the absolute error is $1.96 * ((409.5)^2/1200)^{0.5}$ = 11.8 GPD or 3 percent of the mean. The error of the estimate of savings is almost 109 percent. This example demonstrates that sample sizes that produce precise estimates of the level of water use do not imply that the estimated savings will carry the same precision. The latter is dependent on the magnitude of the difference between the two groups and other factors.

The confidence interval surrounding the projected water savings will be even larger because the forecast variance increased with the length of the forecast. At the minimum, econometric forecasts of water use carry the error of ±15 percent. Again, by disaggregating demand forecasts by geographic and user sectors and incorporating information from various sources one could enhance the precision of the baseline forecast of water use.

## Standardizing Water Savings Estimates

Water planners and administrators must have estimates of reliable water savings as well as costs and potential benefits of conservation programs in order to include such programs in their long-term water supply plans. There is a need for greater focus and standardization of procedures for estimating water savings.

More reliable estimates of savings can be obtained from empirical evaluations of the demand management programs which are designed prior to initiating the program. A proper evaluation plan should make provisions for (1) selection of representative samples of water customers, (2) building databases with observations on water use and other customer characteristics, and (3) statistical estimation and verification of water use relationships for estimating water savings. However, it is important to realize that the science of impact evaluation of water management programs is at a stage of infancy. There is not a single best evaluation design for all programs; many new techniques are currently being tested. It is important to use alternative data and measurement techniques in order to verify and validate the results.

## Long-Term Monitoring

Water conservation planners must have a good understanding of how water is being used in their service area in order to facilitate the development of water demand management programs and to more accurately estimate the potential water savings and cost-effectiveness of such programs. They must be familiar with two very basic characteristics of their service area: (1) water use patterns and trends, and (2) characteristics of the customers. The chapter on forecasting primarily focused on the collection and analysis of water use patterns and trends as derived from water utility records and on the collection of aggregate service area characteristics (such as population, housing stock, and employment). However, for water conservation planning, there is also a need for detailed information about customers' water use characteristics, their knowledge of conservation and water-saving techniques, and their practical behavior toward water conservation.

In order to establish a long-term program for monitoring water use and conservation behavior it is usually necessary to conduct a study of baseline conditions (a baseline study). The purpose of such study is to obtain information on how customers are currently using water and to assess their attitudes toward conservation activities. This objective should not be confused with determining "how much water is used" by customers or customer groups, since these data can be obtained from the water utility records. Rather this study objective focuses on "how water is used" by customers and customer classes.

The types of information collected in a baseline study can be used for a number of purposes. First, the information may be very useful for designing effective demand management programs. For designing long-term water demand management programs, planners must have an understanding of current "baseline" conditions. Second, the types of information collected in the baseline study can be used to evaluate the potential impacts of water demand management measures. The potential water savings associated with various demand management measures cannot be estimated accurately without the knowledge of prevailing habits and patterns of water use. Accurate estimates of potential water savings from demand management measures are important for calculating the benefits and costs of implementing various measures. Therefore, a baseline study can be used to provide information for designing and evaluating potential demand management programs.

Because it is nearly impossible and not cost-effective to obtain information on all customers within a water utility service area, it is essentially necessary to select samples of customers. Sampling has many advantages over a complete enumeration (or inventory) of the population under study. These advantages include reduced cost, greater speed of obtaining information, and a greater scope of information that can be obtained. Once samples of customers are identified, the information sought can be obtained by performing surveys of the sample customers.

The baseline study measures the service area characteristics at only one given point in time. Over time, the characteristics of the service area will change. Furthermore, as demand management programs are implemented, customer attitudes and water-using behavior are also likely to change. Although the baseline study will provide conservation coordinators with a starting point from which to initiate or improve upon water demand management efforts, a long-term monitoring program needs to be implemented to assess how conditions are changing in the service area. A long-term monitoring program includes the periodic assessments of water use and conservation behavior characteristics of the customers of the service area.

The specific objectives of periodic assessments include:

1. To provide information for assessing the market penetration of implemented demand management programs
2. To provide information regarding the satisfaction with implemented demand management programs
3. To provide information for assessing the changing acceptance of demand management programs and devices
4. To provide a means for monitoring changes in customer characteristics, attitudes, and conservation behavior.

## Reference

Kiefer, T. C. and W. Y. Davis. 1991. Tampa Residential Retrofit Evaluation: Analysis of Pilot Program. Carbondale, IL: Planning and Management Consultants, Ltd.

# 11

# Integrating Water Supply and Water Demand Management

**Janice A. Beecher**

*Indiana University–Purdue University, Indianapolis*

The role of the demand side in water utility planning is obvious and important, perhaps more so today than historically. Future water needs must be anticipated and estimated; future utility capacity must be built accordingly. Although demand forecasting and planning always have been intrinsically related, utility planning models and methods have evolved in ways that explicitly recognize the value of demand analysis. The value of these tools has increased in proportion to the increasing value of water resources.

Modern utility planning emphasizes forecasting with greater precision and methodologically sophisticated approaches to risk and uncertainty. But modern planning also embraces the concept of demand *management,* which suggests that water demand is not fixed and unalterable (see Figure 11-1). Water demand is malleable, and manipulating patterns of water use can improve the efficient and effective use of water supply resources.

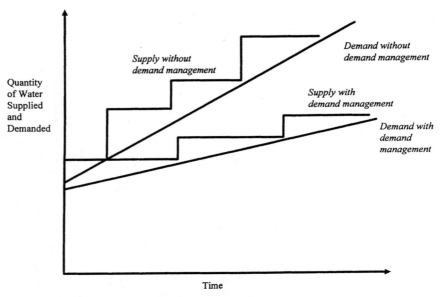

**Figure 11-1.** Projecting Water Supply and Demand with and Without Demand Management

## Integrated Resource Planning (IRP)

The utility planning paradigm perhaps most associated with the concept of demand management is integrated resource planning. *Integrated resource planning* or IRP is a comprehensive form of water utility planning that encompasses least-cost analysis of demand management and supply management options, as well as open and participatory decision making, explicit consideration of risk and uncertainty, and recognition of the multiple institutions concerned with water resources and the competing policy goals among them (AWWA 1994; Beecher, Landers, and Mann 1991; and Beecher 1995). IRP values differ sharply from those characterizing traditional utility planning. A comparison of traditional utility planning and integrated planning appears in Table 11-1.

The concept of integrated planning for water utilities has multiple roots in the multidisciplinary planning literatures. For electric utilities, consideration of demand management was first introduced in the context of least-cost planning (Hanson et al. 1991; Hirst, Goldman, and Hopkins 1990; NARUC 1988). Through the 1970s, inflation, construction-cost overruns (particularly for nuclear power), and poor planning caused energy prices to spiral and led policymakers to question the traditional approach

**Table 11-1.** Traditional Planning and Integrated
Planning Compared

| Criteria | Traditional planning | Integrated planning |
|---|---|---|
| Orientation of the Planning Process | | |
| Resource options | Supply options, little diversity | Supply and demand options, diversity is encouraged |
| Resource evaluation criteria | Maximize reliability, minimize prices | Multiple criteria, including resource diversity, risk management, environmental quality, and public acceptance |
| Resource selection | Based on a commitment to a specific option | Based on the development of a mix of options |
| Resource ownership and control | Centralized, utility-owned | Decentralized, utilities and others |
| Planning Procedures | | |
| Nature of process | Closed, inflexible, internally oriented | Open, flexible, externally oriented |
| Judgment and preferences | Implicit | Explicit |
| Decision tools | Dispute resolution | Consensus-building |
| Stakeholders' identity | Utility and its ratepayers | Multiple interests |
| Stakeholders' role | Disputants | Participants |
| Planning Objectives | | |
| Scope of objectives | Single | Multiple, as determined in the planning process |
| Supply reliability | Constraint | Decision variable |
| Environmental quality | Constraint | Objective |
| Role of pricing | Cost recovery | Economic signal |
| Efficiency | Operational concern | Resource and priority |
| Trade-offs | Hidden | Openly addressed |
| Uncertainty | Uncertainty should be reduced | Uncertainty should be analyzed |
| Risk | Risk should be avoided | Risk should be managed |

SOURCE: Beecher (1995).

to capacity expansion. Analysts discovered that *negawatts* could be less expensive to develop than *megawatts*. Energy-efficient refrigerators, for example, could produce a new "source" of electrical power at a lower cost than building a new power plant. Public policy at the federal and state levels also began to recognize the potential merits of efficiency. Frustration with the concept of "least cost" (that is, least cost to whom?) eventually led to the somewhat broader concept of integrated resource planning. In the mid-1990s, facing increasing competition and fundamental structural change, the energy industry introduced the concept of *distributed resources*, which are smaller, incremental units of supply provided by diverse entities through a variety of supply-side and demand-side technologies.*

Demand management principles were introduced to water-sector planning during the same general time frame, although the term *negagallons* may not have caught on (Baumann, Boland, and Sims 1980, 1981). In many respects, the IRP paradigm is especially suitable to water because it is a natural resource and water services are considered particularly essential. Integrated planning concepts and applications continue to evolve in interesting and important ways. The next iteration by some accounts will be *total water management,* which broadens the spectrum even further to recognize that water utilities are part of a larger water ecology and must be responsible stewards of nature's most precious resource. Making the case for a paradigm shift (let alone pinpointing its occurrence) can be difficult, but the emergence of integrated resource planning in the water sector appears to mark a new way of thinking about how future water needs will be met.

In an integrated framework, a wide range of supply-and-demand management strategies are considered (see Table 11-2). These strategies will vary according to implementation time frame and managing agent (that is, the supplier or the consumer). Demand management can lower system capacity requirements and extend the useful life of water supply, storage, and treatment facilities. Demand management also can make it possible to postpone expensive and "lumpy" (that is, large) capacity additions, such as reservoirs. Finally, demand management lowers operating costs (particularly chemical and energy costs) and the cost of water purchases and water rights. Savings from demand management can be evaluated in terms of *avoided cost,* or the incremental cost associated with not having to produce additional units of water or water service (Beecher 1995, p. 42; Beecher 1996; Gregg and McReynolds 1996, pp. 691–695).

Cost comparisons play a role in virtually every IRP, as seen in the cases summarized in the following sections. Integrated planning not only pro-

---

* The Electric Power Research Institute provides a newsletter and other information materials on distributed resources.

**Table 11-2.** Supply-Management and Demand Management Practices

| Managing agent | Time frame | Supply-management practices | Demand management practices |
|---|---|---|---|
| Water suppliers | Short-term strategies | Supply audits and metering<br>Leak detection and repair<br>Resource management<br>Transfers, diversions, and auxiliary supplies<br>Pressure reduction<br>Relaxation of standards (extreme emergencies only) | Shortage rates, penalties, and surcharges<br>Pleas for voluntary use reduction<br>Use bans and rationing |
| | Long-term strategies | Phased source development<br>Additional storage and conveyance capacity<br>Loss-reduction program<br>Resource management and conjunctive use<br>Transfer, diversions, reallocation<br>Imports (Canada/Mexico)<br>Reclamation and reuse<br>Desalination | Conservation programs for each water use sector<br>Comprehensive metering for all water uses<br>Marginal-cost pricing<br>Water use audits<br>Public information and education<br>Plumbing efficiency standards and retrofits |
| Water consumers | Short-term strategies | Participation in planning and management processes | Reduce use through low-cost technologies<br>Reduce use through behavior and habits<br>Use of substitutes (for nonconsumptive activities) |
| | Long-term strategies | Participation in planning and management processes | Efficient appliances and fixtures, including retrofits<br>Efficient landscaping and irrigation practices<br>Reuse, recycling, and recirculation<br>Agricultural, industrial, and commercial efficiency applications |

SOURCE: Author's construct.

vides a fr                                          ple, cost per million gallc                                          des a framework
for analy                                           nalities (that is,
spillover                                           Jnder traditional
planning                                            ironmental costs
can be qu                                           ternatively, envi-
ronmenta                                            evaluation criteria in scr                                          he method used,
considera                                           rs and decision
makers w                                            urce options.

But int                                             ot an IRP make.
As noted                                            en and participatory. This                                          or style of traditional pla                                          xpand the range
of options considered (a substantive benefit), but also serve to educate
the participants and build consensus for the plan. Participation includes
opportunities for key stakeholders, as well as members of the general
public, to provide input to the water utility. Involving the public at large
and maintaining a level of interest have proven the most challenging (Call
1996; Gardener 1996).

Integrated plans also involve explicit means of considering risk and uncertainty. Understandably, as in the past, the risk of a water shortage usually
tops the list of planner worries. But planners also know that driving risks to
zero usually is cost-prohibitive. An optimal solution is to reduce risk as long
as the marginal benefits of doing so outweigh the marginal costs (Rodrigo,
Blair, Thomas 1995). Risk and uncertainty can be analyzed through qualitative analysis, ranking ordering, probabilistic mapping, and simulation
modeling (Chesnutt and McSpadden 1994). Of course, the longer the planning horizon, the greater will be the uncertainty that accompanies every
option considered. Some of the better techniques for dealing with long-
range uncertainty (such as simulation) will enhance planner perspectives
but may be lost on many decision makers and the public.

Finally, integrated planning also is supposed to consider the multiple
institutions concerned with water policy. This form of integration is perhaps a key distinguishing characteristic of water IRP, as compared with
energy IRP. Essentially, water utilities cannot plan in vacuum. Regulatory
and planning agencies at the federal, interstate, state, and local levels can
and will affect utility resource options. As many of the available case studies illustrate, water utilities are finding it necessary to design solutions on
a regional basis in order to fully address economic, environmental, and
institutional issues. Increasingly, institutional integration will be required
to ensure better planning outcomes (Ruzicka 1996). Although IRP is no
panacea for the water supply industry, it does provide tools for making
difficult choices in a complex world.

## The IRP Process

Various schematics have been put forth to represent the basic steps in the IRP process (see Figure 11-2). The basic steps in integrated planning are roughly similar from process to process. These steps follow a systems approach, meaning that discrete steps in the planning process are highly interrelated and feedback mechanisms are continually at work. IRP is considered highly iterative, meaning that planners will likely run and rerun various models and scenarios, even as some options are actually implemented.

The IRP process can be briefly summarized in seven basic steps (Beecher 1996, p. 44). The first step involves an evaluation of water demand and supply needs. Forecasting techniques, particularly those used by larger systems, have become rather sophisticated. In most cases, alternative forecasts will be developed (for example, high, medium, and low scenarios)

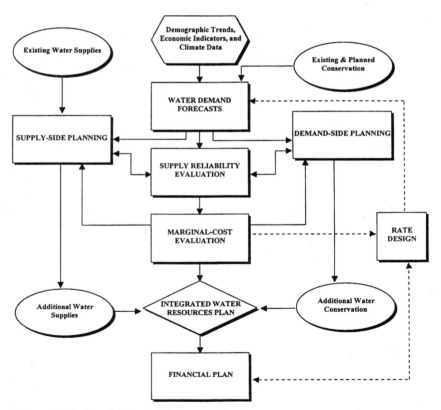

**Figure 11-2.** The Integrated Resource Planning Process as Depicted in Rodrigo, Blair, and Thomas (1995)

based on varying assumptions about key demographic, economic, hydrologic, climate, and other variables. The second step in the IRP process involves determining the principal objectives of the process itself. Objectives should reflect broad philosophical principles and values, but be stated in relatively specific terms. For each objective, specific evaluation criteria should be developed for use throughout the planning exercise. A weighting system can be developed to place greater importance on certain criteria. Third, feasible resource alternatives must be identified. The search can be very broad based, but screening devices can be used to exclude options that blatantly violate one or more evaluation criteria (including, for example, illegal options).

The fourth step involves mixing and matching resource options to construct alternative scenarios, strategies, or portfolios that meet overall resource needs. This part of the process is highly iterative because it involves mixing not only mixing resource options but mixing various assumptions about their availability (for example, probable yields from a supply source or probable participation in a demand management program). The fifth step in IRP involves actually selecting the resource portfolio that best satisfies the initial objectives and evaluation criteria. As noted, the selected mix may not satisfy the criterion of least cost in a strict sense. However, the planning method will provide decision makers with a means of justifying the ultimate selection based on the trade-off between least cost and other criteria. The sixth step is to implement the plan. This involves committing financial and other resources to the preferred alternatives. It also involves coordinating planning activities with other activities within the utility organization (such as financial planning and rate making). Ongoing coordination with outside organizations (such as regulators) also might be required.

The last step in the process concerns evaluation, although it actually is an ongoing process. Evaluation provides a feedback mechanism to decision makers in terms of whether anticipated outcomes actually materialize, and whether unanticipated effects also occur. These simplified steps vastly understate the investment of time and effort devoted to integrated planning. But while some water utilities remain unsure whether they can afford to conduct an IRP, others believe that they cannot afford the alternatives.

## Case Studies in Water IRP

Integrated planning in the water sector is evolving to the point where interesting and important case studies are emerging. No "right" way of integrated planning exists. With each experience, however, more is learned about the planning process and the key ingredients of effective analysis and implementation.

Several brave water utilities have proved to be capable pioneers in the IRP movement. Five recent cases of integrated resource planning are presented here: New York City; Wichita, Kansas; Seattle, Washington; Southern Nevada Water Authority; and Metropolitan Water District of Southern California. The water purveyors in these five regions all share similar concerns about future supply adequacy, reliability, and cost. To varying degrees, each also is concerned about environmental integrity, institutional legitimacy, and public acceptance. The approach taken in each case can readily be classified as an IRP approach, yet each is unique in certain respects. That finding in itself supports the importance of adapting the process to the needs and circumstances of the particular localities engaged in integrated planning rather than assuming that "one size fits all."

## New York City

New York City has been much maligned for public infrastructure issues. In the water sector, the provision of unmetered service is a persistent source of criticism. In the city's defense, of course, the sheer enormity of the city's population means that the challenges faced by New York planners and managers are always on a uniquely large scale.

By the early 1990s, it became obvious that after years of disrepair and neglect, the city's 6000-mile water and wastewater infrastructure badly needed renovation (Ostrega 1994). As demand continued to climb, water supply facilities serving the city repeatedly exceeded safe yields. Wastewater treatment capacity proved to be inadequate as well. Between 1985 and 1993, water rates had more than doubled because of lost governmental subsidies and federal legislation requiring the city to stop dumping sewage sludge in the ocean.

Mounting cost pressures made it necessary for the city to identify its least-cost options and demand management moved quickly to the forefront of the city's efforts. Steven Ostrega, the Deputy Commissioner of the New York City Department of Environmental Protection, refers to water conservation as the city's "new-found religion." A comprehensive set of conservation initiatives, including education, metering, leak detection, and water use regulation were implemented (Table 11-3). Although these efforts began to yield results, stepping up conservation efforts became a priority in the context of long-term resource planning.

The comprehensive list of resource options for New York City is summarized in Table 11-4. The City's ambitious program to replace more than 1 million toilets over a three-year period clearly is one of the most impressive

**Table 11-3.** New York City Water Conservation Initiatives
and Anticipated Savings

| Program | Anticipated water savings |
| --- | --- |
| Universal metering | Incentive-based billing system producing 100 to 150 MGD savings in the previously unmetered residential sector |
| Toilet rebate program | 33 percent of all toilets are replaced, resulting in a savings of approximately 90 MGD |
| Residential water survey program | 6 to 20 MGD savings, with higher savings reflecting leak repairs by owner; provides city with information about the housing stock and elevates conservation consciousness |
| Sonar leak detection | Save approximately 30 MGD and increase crew productivity (increase leaks repaired per day), thereby decreasing leak correction time |
| Hydrant locking devices | Approximate savings of 0.5 MGD in average weather but as much as 100 MGD on days above 90 degrees; reduces public criticism of utility's conservation efforts in an environment of open hydrants |

SOURCE: Ostrega (1994).
MGD = millions of gallons daily
gpm = gallons per minute

examples of modern water demand management (Leibold 1994, 1996).* At the initial level of program funding, 1.25 million toilets could be replaced. The toilet-rebate program aims to accelerate the replacement of water-guzzling toilets (typically 5 gallons per flush) with water-efficient models using only 1.6 gallons per flush (consistent with today's national standards). Replacements began in 1994, beginning in the Bronx and extending in phases to Manhattan, Brooklyn, and throughout the city.

The $240 rebate level was chosen because it would encourage the desired amount of participation at the desired cost. Although this option was the clear favorite in terms of cost, other considerations favored its selection as well. Conventional supply options would require wastewater treatment capacity for the additional sewage flows. The short lead time of the toilet replacement program (three to four years) compared favorably to supply projects requiring ten years or more for planning, design, permit acquisition, potential litigation, and actual construction. Contentious water qual-

---

* This summary is based on data and information handouts, and personal communications in 1994 and 1996 with Warren Liebold, Director of Conservation, Bureau of Water and Energy Conservation, New York City Department of Environmental Protection.

**Table 11-4.** Resource Analysis for New York City[a]

| Resource option[b] | Supply (MGD) | Supply cost ($mil)[c] | Cost of additional wastewater treatment capacity ($mil) | Cost/ MGD capacity ($mil)[c] | Total cost per gallon (cents)[d] |
|---|---|---|---|---|---|
| Current supply and treatment sources | na | na | na | na | 0.189 |
| $150 toilet rebate[e] | 40.2 | $150 | negative | $3.73 | na |
| $240 toilet rebate[e] | 90.6 | $393 | negative | $4.34 | 0.104 |
| $360 toilet rebate[e] | 150.9 | $860 | negative | $5.70 | na |
| Chelsea pumping station expansion[f] | 250.0 | $1200 | $1500 | $10.80 | 0.318 |
| Hudson skimming project | 400–1000 | $4000–8000 | $2500–6000 | $14–16 | na |

SOURCE: Data provided by New York City Department of Environmental Protection, Bureau of Water and Energy Conservation (1994).

na = not applicable

[a] Costs are in present value using an 8 percent discount rate. The consumer energy savings that will result from showerhead replacement are not included in the analysis but are estimated to be $8.4 million annually.

[b] The Chelsea supply expansion project time frame is six to eight years; the time frame for the toilet rebate program is three to four years.

[c] Only capital costs associated with supply and treatment are included in the comparison of capacity costs.

[d] Total capital and operating costs for water supply and wastewater treatment (pollution control) are included in the comparison of cost per gallon.

[e] Rebate program costs include all costs, including rebates, administration, and a private sector program coordinator.

[f] Operating costs, primarily due to filtering, are estimated at approximately $50 million annually.

ity and environmental issues also were significant factors in the analysis. Positive customer impacts weighed in heavily as well. By replacing toilets, metered customers could see reductions in total water and wastewater bills of 20 to 40 percent; bill reductions would be greater if the replaced fixtures were leaking. Furthermore, rate increases would not be needed to cover additional capacity costs.

Metered and unmetered water customers in New York City, including city residents served by the privately owned Jamaica Water Supply Company, were allowed to participate in the toilet rebate program.* Homeowners, apartment-building owners, and commercial-property owners can receive rebates amounting to: (1) up to $240 of the installed cost for the first

* Jamaica serves some customers in the eastern part of Queens. However, New York City provides wastewater service to these customers and supplies some water to Jamaica as well, which is why some Jamaica customers are included in the rebate program.

bathroom in a dwelling unit (defined as a private home or apartment), (2) $150 for each additional bathroom in the same dwelling unit, and (3) $150 for toilets installed in nonresidential buildings. At least 70 percent of the toilets in multifamily or commercial buildings must be replaced for participants to earn a rebate. The city provides an ample list of qualifying fixtures to eligible participants, who must select a licensed master plumber to perform the installation. The program also requires the simultaneous installation of certified showerheads and faucet aerators at the premises. Water savings are assured through professional installation, as well as regulations guiding the disassembly and disposal of the replaced toilet fixtures (which assures that they will not enter secondary markets). Rebates are issued to participants (or to their plumbers at each homeowner's discretion) within thirty days of notification to the program office, subject to spot inspections conducted within that time period. The program is not designed to cover all replacement costs. The payback period, realized through reduced water bills, is expected to be under one year for many water customers.

An important element of the New York toilet rebate program is the partnership with a private contractor (VOLT Information, Energy and Water Technologies, or VIEWtech), and a carefully crafted system of program performance incentives. The contractor helps promote the program, processes applications, performs inspections, and distributes the rebate checks. The contractor is paid on a unit basis, contingent on whether it meets specific performance criteria. By using a private firm, the city is taking advantage of and encouraging the emerging competitive market for conservation services. Private-public partnerships can help water utilities use market forces (such as profits) and market mechanisms (such as competitive bidding), and possibly avoid overbureaucratization of conservation programs. Private vendors also can assume a considerable amount of risk associated with the successful performance of programs, in part because of the opportunity costs associated with failure. Program integrity, monitoring, and evaluation are strongly emphasized by the architects of the New York program. Detailed evaluations will encompass program results in terms of consumer satisfaction, as well as water, wastewater, and energy savings.

By early accounts, New York's toilet rebate program has been very successful. As of the end of March 1996, approximately 775,000 toilets had been installed and as many as 300,000 additional rebate applications had been received. The application process was terminated in mid-April 1996.

## Wichita, Kansas

Wichita initiated an assessment of the community's growing water demand and future supply needs in 1992, which led to an integrated resource planning approach (Klein et al. 1996). The findings indicated that, based on conventional operating practices, existing water supplies were being fully utilized. The investigation of water supply alternatives was broadened to include nonconventional water sources that do not typically have firm yields. According to analysts:

> Because no new conventional water sources were readily available in the surrounding area, the City reexamined how existing regional water resources could be used more effectively through integrated operations to meet projected water demands through 2050. Much developmental work and information gathering needs to be done. For the project to be successful, the City will continue to seek input and support from the community, affected water users, and regulatory agencies (Klein et al. 1996, p. 421).

The Wichita case is noteworthy for its very long term analysis, the number and variety of water resource options considered, and the emphasis on regional coordination issues. The Wichita case is especially useful in recognizing the importance of regulatory institutions in affecting water resource options. These institutional considerations are essential for understanding the true feasibility of implementing a given option. Regulatory considerations included the need to acquire additional water rights, aquifer anti-degradation policies, drinking water quality impacts and compliance with Safe Drinking Water Act provisions, environmental impacts and compliance with federal and state environmental protection and historic preservation laws (Klein et al. 1996).

Analysts in Wichita summarized the key elements of their "customized" integrated plan as follows (Klein et al. 1996, pp. 58–59):

> Implement water conservation to help control customer demand and water use;
> Evaluate existing surface water and groundwater sources to determine their capacity and condition, methods of enhancing their productivity, and ways to protect their quality;
> Evaluate nonconventional water resources for meeting future water needs;
> Optimize all available water resources to enhance water supply;
> Pursue an application for conjunctive water resource use permit from state agencies;
> Evaluate the effects of using different water resources of water supply, delivery, and treatment facilities with consideration of risk and reliability; and
> Communicate with key stakeholders including regulatory agencies, other water users, and the public.

The analysis of resource options for Wichita was comprehensive and straightforward. A large matrix analysis was used to display 27 conventional and nonconventional resource options and their key characteristics (Warren et al. 1995, pp. 57–71). A select number of variables extracted from the matrix appear in Table 11-5. For each option, the analysis considered: construction costs, expected available flow (including alternative scenarios when applicable), unit costs, general advantages and disadvantages, and specific implementation issues related to policy or political, legal, environmental, and water quality concerns. A screening process was used to eliminate several options from further consideration. For example, the "no action" alternative was dropped "because it would limit growth and risk loss of future development, tax revenues, and jobs" (Warren et al. 1995, p. 64). The remaining options were ranked in terms of overall desirability.

Planners in Wichita recognized that water supply operations are growing in complexity and that operational trade-offs are necessary when implementing an integrated approach. The key benefit to better planning, however, is the more effective use of the region's water resources.

### Seattle, Washington

Seattle's rainy reputation belies the region's water needs, particularly during the summer and fall months. Seattle water planners recognize that water resources are as important to the aquatic species of the region as to the human variety. The Cedar River supports the nation's largest sockeye salmon run, making it necessary to manage the region's water resources for multiple uses (Seattle Water Department 1992). Many of the lessons learned in Seattle are relevant for other communities engaged in integrated planning.

The Seattle Water Department's Comprehensive Regional Water Supply Plan was published in 1992 and approved by the city council in late 1993 (Drury 1996, pp. 687–690). The plan, spanning a 30-year planning horizon, represents a five-year investment in integrated resource planning for the largest water supply utility in the Pacific Northwest. The system provides retail service to a population of nearly 600,000 and wholesale service to an additional 670,000. Regional cooperation and maintaining the environmental integrity of the region are heavily emphasized in the plan. The focus on "broader policy choices and action strategies" in three major areas—conservation, supply development, and regional cooperation—clearly distinguish it from "a typical utility plan" (Seattle Water Department 1992, p. 4).

Seattle's approach is philosophical and reflective of the system's long history of growth and change. Guiding policies used to develop the plan were organized into two broad categories. "Guiding policies" were designed to reflect fundamental values: public health and safety, service excellence, environmental stewardship, economics, regional coordination, and workforce diversity. "Action policies" were designed to guide specific operational areas: level of service, planning and service areas, service

**Table 11-5.** Resource Analysis for Wichita, Kansas

| Resource alternative | Expected yield (MGD) | Construction cost ($mil) | Unit cost ($/mil. gal.) | Rank[a] |
|---|---|---|---|---|
| Kanapolis Reservoir (existing) | 10 | 69 | 344 | ns |
| Milford Reservoir (existing) | 60 | 155 | 141 | 9 |
| Corbin Reservoir (proposed) | 35 to 53 | 470 | 669 | ns |
| Douglass Reservoir (proposed) | 14.2 | 202 | 707 | ns |
| Murdock Reservoir (proposed) | 35 | 231 | 329 | ns |
| Equis Beds: purchase water rights | As available | $400/acre-ft | 1,227 | 8 |
| Equis Beds: Burton special use and control areas | 9.8 | 26 | 130 | ns |
| Haysville Groundwater | 2.85 | 22 | 386 | ns |
| Reserve Wellfield | 10.8 | 1.0 | 4.7 | 6 |
| Reserve Wellfield (peak-use only) | 10.8 | 1.0 | 37 | 6 |
| Gilbert-Mosley remediated groundwater | 3 | 1.5 | 25 | 4 |
| Arkansas River supply to water treatment plant | 0 to 8 | 21 | 132 | ns |
| Little Arkansas River supply to water treatment plant | 0 to 44 | 21 | 23 | 2 |
| Cheney Reservoir: operations modifications | up to 60 | 0 | 0 | 5 |
| Cheney Reservoir: purchase flood storage | 3 | na | na | [b] |
| Membrane filtration plant | 60 | 34 | 168 | ns |
| Membrane filtration plant | 60 | 191 | 158 | ns |
| Cheney overflow pipeline to water treatment plant | 28 | 53 | 96 | 7 |
| Cheney overflow pipeline to water treatment plant | 35 | 60 | 87 | 7 |
| Cheney overflow: side storage reservoir | 28 to 35 | na | na | ns |
| Cheney overflow: subsurface storage | 34 | 65 to 165 | 94 to 237 | 10 |
| Little Arkansas River: subsurface storage | 34 | 26 to 126 | 46 to 219 | 3A |
| Treated wastewater reuse: local irrigation | 1.1 | 15 | 1,336 | 11 |
| Treated wastewater reuse: subsurface storage | 68 | 130 to 230 | 96 to 169 | ns |
| Treated wastewater reuse: sell to irrigators | 68 | 129 | 95 | ns |
| Little Arkansas River: bank storage | 7 to 39 | 6.2 to 175 | 45 to 221 | 3B[b] |
| Little Arkansas River: bank storage | 7 to 39 | 11.5 to 164 | 41 to 207 | 3B[b] |
| Rain harvesting | .007/unit | 0.6/unit | 4,117 | ns |
| Excess potable water production: subsurface storage | na | na | na | ns |
| Low-range water conservation | 15 | 23 | 77 | 1 |
| No action | 23 | 0 | 0 | ns |

SOURCE: Warren (1995).

na = not applicable

ns = not selected as a viable alternative based on screening level cost

[a] Rankings were based on a variety of criteria, including but not limited to the cost criteria provided.

[b] Further analysis needed to confirm viability.

to customers, resource selection, providing high-quality drinking water, maintaining system facilities, responsible land management, education and community relations, and information management. Each policy is accompanied by a statement of philosophy and intent, as well as guidelines for implementation.

Eighty percent of Seattle residents surveyed agreed with the statement, "It is important to conserve even when there isn't a water shortage" (Seattle Water Department 1992, p. 7). This strong conservation ethic is reflected in the plan's commitment to programmatic conservation as a resource option:

> "Resources" are often thought of in terms of additional supply. However, making more efficient use of an existing supply also creates a resource. While the department's sources of supply are central to meeting demand, conservation is a real "source," too—and in many ways a very attractive one. Conservation is the source of additional water we need in the 1990s. It is less expensive and has fewer environmental impacts than supply options. It's the best fit for our water supply problem: summer season demand (Seattle Water Department 1992, p. 4).

Seattle's comprehensive plan includes a comprehensive approach to conservation, including retrofits, audits, regulations, and seasonal pricing to suppress peak-season demand. A summary analysis of the plan appear in Table 11-6. Many elements of the city's program will accelerate customer investments in conservation that might occur "naturally" in response to price changes and other factors. Beyond conservation, alternative supply options were evaluated in terms of cost effectiveness, environmental sensitivity, and water system reliability. Supply options were narrowed from 22 to six. The water department also is implementing "parallel planning," by which it can begin developing the preferred supply option, while keeping supply alternatives open to further consideration as conditions change.

Seattle's planning process was patterned after the model integrated approach, including least-cost analysis, open and participatory decision making, and consideration of the multiple institutions concerned with water policy (Drury 1996). The reality of the planning process, however, did not always live up to expectations. First, although conservation presents a least-cost and less-uncertain alternative in the short term, supply development probably will be needed. To planners, the benefits of conservation at times appeared "short-lived" when compared to long-term supplies.* Second, involving the public in planning is more difficult in practice than in theory. Ironically, the public may be more attentive to the

---

* Of course, the benefits of end-use efficiency will extend into the future, extending the useful life of future supplies as well as existing ones.

**Table 11-6.** Resource Analysis for Seattle, Washington

| Resource | 1990 cost ($ millions) | Estimated yield (MGD) |
|---|---|---|
| Programmatic conservation<br>  Comprehensive showerhead program<br>  Residential water audits (peak)<br>  Nonresidential (commercial) water audits<br>  Plumbing code trade allies program<br>  Outdoor water regulations<br>  Irrigation audits/evapotranspiration<br>    stations<br>  Nonrevenue water reductions<br>  Regional toilet retrofit<br>  Commercial toilet retrofit | $15.9 | 12 MGD by 2002 |
| Pricing: Full marginal-cost summer rates | None | 11 MGD by 1002<br>21 MGD by 2020 |
| Plumbing code | None | 9 MGD by 2002 |
| New supply: North Fork Tolt Diversion | 27.1 to 100 | 52 MGD |
| New supply: Morse Lake Permanent<br>  Pumping Plant | 73.6 | 44 MGD |

SOURCE: City of Seattle Water Department, *Seattle Water Department Water Supply Plan* (Seattle, WA: Seattle Water Department, July 1992). *Highlights,* table 11-1.

*outputs* of the traditional planning process (that is, pronouncements of new supply projects) than to providing *input* in the context of a participatory approach. Third, policy and technical advisory committees greatly broadened the planning perspective. However, their interests and motivations were not necessarily matched by the elected officials responsible for approving the plan. Finally, the planning process occurred over a very lengthy time period. Participant turnover and revisiting of issues though resolved were problems. Nonetheless, despite these shortcomings, analysts found integrated planning to be appropriate and necessary.

### Southern Nevada Water Authority

The integrated planning process designed for the Southern Nevada Water Authority used a probabilistic approach to long-term, regional water supply issues. Architects of the Southern Nevada method stressed the importance of adapting the approach to local conditions, involving the public in decision making, and approaching planning as a dynamic and nonlinear process (Fiske and Dong 1995, pp. 72–83).

The Southern Nevada Water Authority signaled its intention to conduct an integrated resource plan in 1994. The Authority is a consortium of water supply and wastewater treatment providers serving the rapidly growing and very arid vicinity around Las Vegas, Nevada. The integrated plan will guide the Authority's resource decisions through the year 2050.

Several major components of the integrated planning process are well illustrated by the Southern Nevada experience (Fiske and Dong 1995). First is the development of demand forecasts. In this particular case, based on the high-demand forecast, available permanent and temporary water supplies will be exceeded approximately in the year 2025. Given this scenario, water agencies and other stakeholders should develop explicit policy objectives to guide the analysis of resource options. These objectives can cover a range of issues and concerns, including the traditional objective of maintaining safe and reliable water supplies, but also concerns about the human and natural environments. In addition to stating each objective in precise terms, planners also should develop measurable evaluation criteria for each objective. For many types of objectives, an ordinal ranking can be used (for example, a 10-point scale indicating how well a strategy meets the objective).

In the next part of the process, supply-side and demand-side options for meeting future needs are examined. Supply options can be evaluated in terms of such issues as hydrology and water rights, environmental impact, water quality and treatment requirements, facility siting, transmission options, costs, and public acceptance (Fiske and Dong 1995, p. 77). Demand options (on the "customer's side of the meter") present a unique assessment challenge. Conservation measures can be passed through qualitative and economic screens and ultimately grouped into programs, which in turn can be compared with supply options in terms of meeting agency objectives. The evaluation process is complicated, but greatly enhanced, by expressly considering uncertainty. In the Nevada case, this was accomplished by assigning probabilities to the various resource options and displaying them in a series of decision trees.*

Based on this information, resource sequences and strategies can be developed. According to Fiske and Dong (1995, p. 78):

> A sequence is defined as a deterministic progression of utility actions regarding facility and (demand-side and supply-side) resource addi-

---

* Three-point probability distributions were used. For a given option, for example, analysts might project an 80 percent probability of realizing 100,000 acre-feet of capacity; a 10 percent probability of realizing a 70,000 acre-feet of capacity; and a 10 percent probability of realizing 130,000 acre-feet of capacity. Each probability appears on its own "branch" in the decision tree. A drawback of decision-tree analysis in planning is that with multiple options the method quickly becomes very cumbersome. In addition, a multitude of discrete probabilities can lend an aura of precision to a rather imprecise exercise.

tions. This set of actions presumes a particular set of uncertainties and their outcomes.

A strategy is defined as a probabilistic multibranched "tree" of sequences defining the utility actions that should be taken under various sets of uncertainties and their outcomes.

Probabilities are used to construct alternative sequences and a sequence can be eliminated if it is clearly inappropriate (for example, if it does not result in supplies sufficient to meet demand). In the final phases of the analysis, the alternative strategies are evaluated against the basic evaluation criteria and a strategy or strategies are selected. Short-term and long-term options can be combined. A strategy evaluation matrix is used to present decision makers with a summary analysis where the trade-offs among alternative strategies (for example, cost and environmental impact) are easily seen. A sample matrix is presented in Table 11-7. After an institutional analysis, the resource plan can be formulated and implemented.

The Southern Nevada planning process was under way at the time of this writing. The first phase, which involved developing forecasts, identifying objectives and evaluation criteria, and defining uncertainties, took approximately one year to complete. As in other cases of planning, the task of the analysts was made more difficult by data weaknesses and institutional constraints on supply options involving the Colorado River. The options under consideration include combinations of contingency and planned conservation, reuse, seasonable groundwater recharge, and development of new transmission and treatment facilities.

## Metropolitan Water District of Southern California

California has long contended with the hydrological, sociological, and political dimensions of water supply. The region's abundant sunshine and temperate climate are not matched by abundant water to satisfy the needs of a burgeoning population. By some expectations, even the existing supplies on which Southern California depends could diminish. Alternative supply technologies, including long-distance pipelines and desalination, seem to first appear on the West Coast. Although some believe that California water problems are unique, others believe that problems first experienced in California eventually will beset other regions of the country.

The Metropolitan Water District of Southern California, or Metropolitan, essentially is a water "importer" serving 27 local water agencies in the region through wholesale contracts but also through support and assistance for developing local water supply resources (including conservation). The

**Table 11-7.** Resource Analysis for the Southern Nevada Water Authority

| Strategy | Expected cost ($mil.) | Reliability | | Human environment | | Natural environment | | Public acceptability |
|---|---|---|---|---|---|---|---|---|
| | | Probability of shortage for the year 2050 | Maximum expected shortage | Cultural and historic | Land use | Aquatic | Terrestrial | |
| A | 502 | 81 | 10 | 3 | 3 | 3 | 3 | 4 |
| B | 580 | 16 | 2 | 4 | 4 | 4 | 4 | 1 |
| C | 644 | 0 | 0 | 5 | 5 | 5 | 5 | 3 |
| D | 634 | 4 | <2 | 5 | 4 | 5 | 4 | 3 |
| E | 602 | 32 | 5 | 4 | 4 | 4 | 4 | 2 |
| F | 671 | 0 | 0 | 5 | 5 | 5 | 5 | 3 |

SOURCE: Gary Fiske and Anh Dong, "IRP: A Case Study from Nevada," *Journal American Water Works Association* 87, no. 6 (June 1995): 72–83.

agency's guiding mission has been "to fully meet all regional water needs beyond those met through local supplies by acquiring new important water supplies and [constructing] transmission, storage, and treatment facilities" (Rodrigo, Blair, and Thomas 1995). Metropolitan began to reevaluate its mission in light of increased uncertainty associated with economic growth, severe drought, court decisions regarding water allocation, environmental regulations governing the State Water Project, water quality concerns, and a planned 10-year, $6 billion capital improvement program (Rodrigo, Blair, and Thomas 1995).

Metropolitan engaged in a comprehensive IRP effort over more than a two-year period beginning in 1993. Metropolitan's planning process targeted two fundamental issues relevant to many jurisdictions. The first was the lack of coordination among Metropolitan and the local water agencies it served. As each agency faced its own cost and resource pressures, each developed its own independent strategy. As is true in many parts of the country, the fragmented nature of planning only threatened to exacerbate the resource problems of the region. A coordinated approach would recognize the interdependence of the agencies and promote more effective resource management (Rodrigo, Blair, and Thomas 1995). The Metropolitan planning process emphasized participation by member agencies, local water retailers, and groundwater purveyors in the region. A work group served as the steering body for the process, augmented by public forums, focus groups, and regional assemblies (MWDSC 1994a and 1994b; Rodrigo, Dickinson, Brown 1996). Thus, the process of planning itself had a fundamental purpose of involving numerous stakeholders.

The second key issue addressed by Metropolitan was supply reliability, another issue that has universal significance. As mentioned earlier, most water providers place a substantial emphasis on reliability. Planners explicitly recognized the trade-off between marginal reliability and marginal resource development costs. Comparing marginal benefits to marginal costs suggests that an optimal level of reliability is preferable to absolute reliability. Metropolitan began its planning process by developing and approving a wholesale reliability goal, which also was translated into a retail reliability goal. But the planning process was used to continually assess the assumptions underlying the reliability goal and the trade-offs required to meet it. Thus, least-cost reliability played an important role in Metropolitan's analytical and resource selection processes.

Metropolitan's planning process involved six steps (Rodrigo, Blair, and Thomas 1995):

1. Develop objectives.

2. Develop evaluation criteria related to each objective.

3. Identify all possible resource options for meeting the objectives, including conservation "best management practices."

4. Develop compatible resource combinations into overall strategies.

5. Evaluate alternative strategies in meeting objectives.

6. Iterate as necessary.

Resource mixes were evaluated according to five criteria: supply reliability, cost, water quality, flexibility and diversity, and institutional and environmental constraints. The analysis also incorporated alternative hydrologic scenarios. Risk was incorporated by limiting the anticipated availability of the option or increasing its cost.

Analysts arrived at a unit cost estimate for each resource option, ranging from under $400 per acre-foot (for additional imports, groundwater storage, conservation, and reclamation) to approximately $1500 per acre-foot (for salination). Analysts also presented the data in terms of investment levels (see Table 11-8). The preferred resource mix eventually supported by the plan did not meet the strict criterion of least cost, but it does represent an optimal combination that will meet supply reliability, water quality, and rate-impact goals in a cost-effective way (Rodrigo, Dickinson, Brown 1996, pp. 505–509). As a result, Metropolitan's diversified resource portfolio for the year 2020 includes: conservation (22%), State Water Project (22%), Colorado River Aqueduct (18%), storage and transfers (16%), groundwater storage (13%), and recycling and reuse (10%). The plan is considered flexible enough to meet changing conditions and needs.

Analysts of the Metropolitan case indicate that integrated planning already has produced three salient results (Rodrigo, Dickinson, Brown 1996, p. 509). First, Metropolitan was able to reduce the cost estimate of its capital improvement plan by 30 percent (from $6 billion to $4.3 billion). The $1.5 million invested on the analytical component of the planning process appears to have been a sound investment. Another much-heralded benefit was the consensus agreement reached on managing the Bay-Delta estuary, after years of controversy and conflict. Finally, the Metropolitan case illustrates the benefits of improved awareness and co-ordination among regional water resource stakeholders.

## Conclusion

These cases are illustrative, but not fully representative, of integrated resource planning in the water sector. Certainly the particular cases are very helpful in understanding the increasing importance of demand forecasting and demand management in utility planning. With each IRP,

**Table 11-8.** Resource Analysis for Metropolitan Water District of Southern California

| Supply source | Investment level 1[a] | Investment level 2[b] | Investment level 3[c] | Investment level 4[c] |
|---|---|---|---|---|
| **Imported Supplies** | | | | |
| Colorado River Aqueduct | Existing entitlement 0.6 maf | Land fallowing and conservation 0.9 to 1.2 maf | Full aqueduct capacity 1.2 maf | na |
| State Water Project | Existing facilities, reduced deliveries 0.1 to 1.0 maf | Interim Delta Project 0.3 to 1.4 maf | Delta Water Transfer Facilities 0.6 to 2.0 maf | South of Delta Storage 1.3 to 2.0 maf |
| Transfers from the Central Valley | Options and spot transfers | na | na | na |
| **Local Supplies** | | | | |
| Local surface and ground waters | Existing Production 1.4 maf | na | na | na |
| Water conservation | Existing conservation 0.25 to 0.30 maf | Fully implemented BMPs 0.70 to 0.75 maf | Aggressive conservation 0.85 to 0.90 maf | na |
| Water reclamation | Existing facilities under construction 0.3 maf | Facilities under design 0.5 maf | Planned facilities 0.7 to 0.8 maf | na |
| Los Angeles Aqueduct | Existing facilities 0.03 | na | na | na |
| Groundwater recovery | Existing facilities 0.03 maf | Planned projects 0.10 maf | Potential projects 0.16 maf | na |
| Ocean desalination | Small facilities 0.01 to 0.03 maf | Mid-sized facilities 0.06 to 0.10 maf | Large regional facilities 0.1 to 0.2 maf | na |

SOURCE: MWDSC (1994, table ES-1).

maf = million acre-feet

na = not applicable

[a] Supplies than can be provided by infrastructure and resource programs than currently exist or are already under construction.

[b] Supplies resulting from projects and programs that agencies have taken steps to implement (such as projects under design), but for which construction has not yet begun.

[c] Potential supplies only in the "concept" phase of development.

demand management options win further credibility as a resource alternative. As experience with IRP grows, more will be learned about the ups and downs of taking a more integrated approach to resource planning.

Not only will IRP continue to evolve, so will the context in which it takes place. Structural change, competition, and rising costs in the water sector will present new challenges to water suppliers. The regulatory environment will continue to place new demands on utilities. Customers will continue to hold water suppliers accountable for safe and reliable water at an affordable price. Water utilities may find that integrated planning provides a more open and comprehensive forum in which to explore complex alternatives and make difficult but necessary choices.

# References

American Water Works Association (AWWA). 1994. *Integrated Resource Planning: A Balanced Approach to Decision Making.* Denver, CO: American Water Works Association.

Baumann, Duane D., John J. Boland, and John H. Sims. 1980. *The Evaluation of Water Conservation for Municipal and Industrial Water Supply: Procedures Manual.* Carbondale, IL: Planning and Management Consultants, Ltd.

Baumann, Duane D., et al. 1981. *Planning and Evaluating Water Conservation Measures.* Chicago, American Public Works Association.

Beecher, Janice A. 1996. "Avoided Cost: An Essential Concept for Integrated Resource Planning," *Water Resources Update* no. 104 (Summer): 28–35.

Beecher, Janice A. 1995. "Integrated Resource Planning Fundamentals," *Journal American Water Works Association* 87, no. 6 (June): 34–48.

Beecher, Janice A., James R. Landers, and Patrick C. Mann. 1991. *Integrated Resource Planning for Water Utilities.* Columbus, OH: The National Regulatory Research Institute.

Call, Chris. 1996. "Demand Management as a Component of IRP: The Long and Winding Road," and Elizabeth V. Gardener, "Denver Water's IRP Process: Getting Buy-In from the Public for Conservation," *Proceedings of Conserv96.* Denver, CO: American Water Works Association: 423–427.

Chesnutt, Thomas W. and Casey N. McSpadden. 1994. *Putting the Pieces Together: Decision Support for Integrated Resources Planning Using IRPSIM.* Santa Monica, CA: A&N Technical Services, Inc.

Drury, Kim. 1996. "Integrated Resource Planning: Experience at Seattle Water Department," *Proceedings of Conserv96.* Denver, CO: American Water Works Association: 687–690.

Fiske, Gary and Anh Dong. 1995. "IRP: A Case Study from Nevada," *Journal American Water Works Association* 87, no. 6 (June 1995): 72–83.

Gregg Tony and Maureen McReynolds, "Austin's Integrated Water Resource Planning Process," *Proceedings of Conserv96.* Denver, CO: American Water Works Association: 691–695.

Hanson, Mark, Stephen Kidwell, Dennis Ray, and Rodney Stevenson, 1991. "Electric Utility Least-Cost Planning," *Journal of the American Planning Association* 57, no. 1 (Winter).

Hirst, Eric, Charles Goldman, and Mary Ellen Hopkins. 1990. "Integrated Resource Planning for Electric and Gas Utilities," a paper presented at the conference on Energy Efficiency in Buildings sponsored by the American Council for an Energy-Efficient Economy (August).

Klein, Jeff, Frank Shorney, Fred Pinkney, Rick Bair, David Warren, and Jerry Blain. 1996. "Integrated Resource Planning at Wichita, Kansas: Addressing Regulatory Requirements," *Proceedings of Conserv96*. Denver, CO: American Water Works Association: 417–421.

Metropolitan Water District of Southern California (MWDSC). 1994a. *Integrated Resources Plan Draft Interim Report* (March 23).

Metropolitan Water District of Southern California (MWDSC). 1994b. "Integrated Resources Plan Assembly: Assembly Statement," San Pedro, California (June 9–11).

National Association of Regulatory Utility Commissioners (NARUC). 1988. *Least-Cost Utility Planning Handbook for Public Utility Commissioners, volume 1*. Washington, DC: National Association of Regulatory Utility Commissioners.

Ostrega, Steven F. 1994. "New York City: Where Conservation, Rate Relief and Environmental Policy Meet," a paper presented at the Annual Meeting of the American Water Works Association in New York.

Rodrigo, Dan, Timothy Blair, and Brian Thomas. 1995. "Integrated Resources Planning and Reliability Analysis: A Case Study of the Metropolitan Water District of Southern California" (unpublished).

Rodrigo, Dan, Mary Ann Dickinson, and Paul Brown. 1996. "Southern California's Landmark Integrated Resources Plan," *Proceedings of Conserv96*. Denver, CO: American Water Works Association: 195–199.

Rose, Kenneth, Paul Centolella, and Benjamin Hobbes. 1994. *Public Utility Commission Treatment of Environmental Externalities*. Columbus, OH: National Regulatory Research Institute.

Ruzicka, Denise and Bob Hartman. 1996. "Integrated Resources Planning: The Connecticut Experience," *Proceedings of Conserv96*. Denver, CO: American Water Works Association: 411–415.

Seattle Water Department. 1992. *Seattle Water Department Water Supply Plan: Comprehensive Regional Water Supply Plan Highlights*. Seattle: City of Seattle Water Department (July 1992).

Warren, David R., Gerald T. Blain, Frank L. Shorney, and L. Jeffrey Klein. 1995. "IRP: A Case Study From Kansas," *Journal American Water Works Association* 87, no. 6 (June 1995): 57–71.

# 12

# Application of Integrated Resource Planning Approach to Urban Drought

**Benedykt Dziegielewski**
*Southern Illinois University at Carbondale*

Adequate community water supply is an essential service which promotes public health and safety, economic activity, and general community well-being. During the 1980s, hundreds of municipalities across the United States had their water supplies threatened by drought. Many more communities may experience water shortages during future droughts because of increasing demands for water and reduced opportunities for the development of new water supplies. In the past, rapid urban growth has made it possible to design and build water facilities with substantial extra capacity to accommodate population growth and industrial development. Water supply systems which rely on surface water supplies were protected from droughts by providing sufficient storage of water in times of high rainfall for use during periods of drought. Although the provision of extra storage capacity remains one of the most popular means of protection against drought, several new economic, social, and environmental considerations place this alternative beyond the reach of many water agencies.

## Droughts and Water Management

New strategies for protecting urban water supplies against drought are needed in order to overcome a number of limitations on the traditional approaches. First, there is a limited availability of untapped sources of supply. Many urban areas, especially in the West, have begun to experience water allocation problems as regional surface supplies have become fully appropriated and groundwater aquifers have become depleted. Acute or chronic source contamination, particularly of groundwater sources, further limits water availability. Also, large-scale water transfers between river basins or across political boundaries are often not feasible due to legal, political, and environmental constraints. Second, increasingly stringent water purity standards have led to a significant increase in the cost of water treatment and in some cases water sources that served communities for decades are no longer adequate because of excessive contamination. Third, the prospects for financing major construction programs are discouraging for many public utilities. Water supply often competes for funds with other essential municipal services, yet is at a disadvantage because of its high investment requirements and traditionally low revenues due to subsidized pricing. Fourth, new environmental legislation has severely constrained the opportunities and alternatives in urban water supply. Water supply development has to be coordinated with wastewater planning and any major construction of water facilities is subject to extensive review and regulation. Finally, the increasing general concern for environmental quality has resulted in a more active public role in resource management decisions. The need for new supply development receives unprecedented scrutiny from environmental groups and even projects that are partially completed may be halted because of their potential environmental impacts.

These limitations forced water planners to look for new approaches for dealing with droughts. The integrated resources planning process offers a balanced approach which combines both long-term and short-term measures for ensuring the reliability of urban water supply. Under this approach, the best drought plan is one that uses an optimal mix of long-term and short-term options. Possible options under each category are described here.

## Long-Term Protection Against Droughts

In terms of long-term drought protection, a drought that would result in significant supply shortages would have to be at least as severe as the drought of record. Droughts of record are usually of unknown probability of occurrence; however, they can be placed between the recurrence intervals of 50 and 100 years. Dziegielewski et al. (1993) reported that approximately one-half of urban water providers view their systems as vulnerable to drought and are likely to respond to mild or moderate droughts. These respondents seem to place little confidence in the protection built into the concept of design drought.

The overall vulnerability of water supply systems to the adverse impacts of drought can be assessed based on (1) the types of water supply sources and entitlements to water from these sources and (2) the degree of long-term drought protection as indicated by the severity of drought that would cause significant water shortages.

Generally, groundwater supplies are less susceptible to drought than river intakes or surface-water impoundments. However, the susceptibility of both surface-water and groundwater systems may be heightened by (1) the lack of knowledge about the nature of their legal entitlements to water from these sources, (2) the basis on which the quantity of water they are entitled to was established, and (3) the degree to which their water supply would be reduced in the event of a water shortage. Uncertainty about entitlements is usually high in the purely riparian states.

Because of these uncertainties and the limitations on traditional approaches to long-term drought protection, many water agencies are forced to consider demand management alternatives aimed both at controlling growth in water demand over time and at achieving significant temporary reductions in water use during periods of drought. The demand management options that can substantially reduce future water use and enhance the level of drought protection may include:

1. Public campaigns to educate consumers on how to modify water use habits to reduce water consumption

2. Promotion or mandatory requirement of the use of water-saving devices and appliances

3. Promotion or mandatory requirement of low-water-using urban landscaping

4. Adoption of efficient marginal cost pricing strategies to discourage inefficient uses of urban water

5. Adoption of zoning and growth policies to control the number of water users served by the water supply system

These long-term conservation alternatives in combination with some unconventional supply augmentation alternatives have the potential for providing adequate future water supply for urban areas at the minimum cost while enhancing the ability to withstand the effects of droughts.

## Short-Term Drought Management

The limited reliance on long-term drought protection and the readiness to respond to drought events less severe than the design drought requires some level of local drought preparedness. Many water agencies have a specific drought contingency plan in place or they rely on water conservation ordinances that have been adopted in the communities they serve. Finally, water utilities have agreements to rent or purchase water temporarily. Systems with drought contingency plans, water rental agreements and water shortage ordinances represent a way of formalizing drought management activities. Water systems without such plans are likely to resort to ad hoc drought response measures.

During periods of drought, water agencies that must carry the burden of responsibility for uninterrupted supply can take a number of emergency actions in order to minimize the risk of running out of water. Such drought response actions (or drought management measures) may be (1) planned prior to the onset of drought or developed ad hoc after a drought becomes apparent, and (2) oriented toward reducing water demand or toward provision of supplemental (emergency) sources of supply. These actions, whether aimed at increasing supply or reducing demand, may result in a temporary increase in the cost of water supply or may cause the urban economies and consumers to suffer significant economic losses.

In order to effectively manage a drought event, water supply agencies have to make decisions under a great deal of uncertainty. The source of uncertainty comes from the lack of definitive answers to the following questions:

1. Are we in a drought?
2. How long and severe will the drought be?
3. What can we do to avoid water shortages?
4. When should we implement the emergency actions?
5. When should we call off the emergency measures?

These and other questions must be answered before specific drought-response actions are taken. Because the information and data needed to provide definitive answers usually are not available, water managers may choose to wait and monitor the development of drought, thus possibly forgoing the opportunity to conserve the available supply.

Once the drought is recognized and the need to take response actions is no longer questioned, the next problem is what actions should be taken. The range of possible drought management measures is affected by many factors. Many options are precluded because their implementation lead time may be too long, or they may be technically infeasible. Other measures may not be acceptable to the community, or may be too costly. Finally, as with most management decisions, there may be a question of the legality of water agency actions.

## Proactive Drought Management

The adverse impacts of drought on public water supply systems can be substantially reduced by proactive drought management (Moreau and Little 1989), and the development of drought contingency plans can reduce the uncertainty in making drought-response decisions. Such plans can provide guidelines to decision makers as to the proper times to employ drought response measures and which measures to employ depending on the status of demand and supply. However, most appropriate drought contingency plans, have to be developed as a part of a comprehensive drought planning effort where the long-term and short-term drought-management options are considered. A planning framework for finding an optimal balance of long-term and short-term drought protection measures is described next.

## A Drought Planning Framework: DROPS

The steps that need to be taken while developing a plan for coping with potential water supply deficits during periods of drought include: (1) assessment of the reliability of current and new sources of water supply; (2) evaluation of short-term drought-response options; (3) evaluation of selected long-term drought-protection investments; and (4) selection of the optimal drought plan.

The main goal of the drought planning process is to identify the optimal components of a drought mitigation strategy by selecting from a number of alternative strategies aimed at dealing with the risk of water shortages. This can be accomplished using a systematic procedure referred to as the DROPS method (DRought Optimization ProcedureS). A complete elaboration of this method is given in Dziegielewski et al. (1983) and Dziegielewski and Crews (1986).

### The DROPS Approach

The development of drought contingency plans must begin with the determination of the vulnerability of a water supply system to water shortages during drought and an assessment of the economic consequences of potential supply shortages. The magnitude of economic damages during a severe drought should dictate the need for and the level of long-term drought protection. Economic theory suggests that the incremental cost of long-term drought protection should be balanced with the benefits of reducing economic damages that may result from recurring shortages of water.

Four types of adjustments to drought can be distinguished (Figure 12-1). In the long term, the vulnerability of a system to drought could be reduced by augmenting supplies or implementing conservation measures to permanently reduce the demand. A system that has to resort to emergency measures fairly frequently can probably mitigate droughts more efficiently by expanding supply or reducing demand through long-term conservation programs. However, if the risk of drought damages during a planning period is low or moderate, the optimal strategy may involve the formulation of drought contingency plans to cope with actual emergencies. Temporary reduction of demand and reliance on emergency water supplies are two types of short-term adjustments that can be incorporated into drought contingency plans. It is important to note that the "optimal drought plan" can be comprised of the combination of all four types of adjustments shown in Figure 12-1.

The most critical component of the DROPS method is the analysis which leads to the selection of strategies from the four major categories of adjustments to drought. This selection is determined by balancing the incremental cost of the long-term adjustments with the decremental coping cost associated with the implementation of drought contingency plans.

The major steps of the DROPS method are outlined in Figure 12-2. First, the ability of the existing system to support the established uses of water

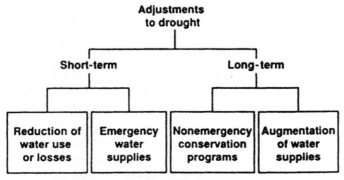

**Figure 12-1.** Types of Adjustment to Urban Drought

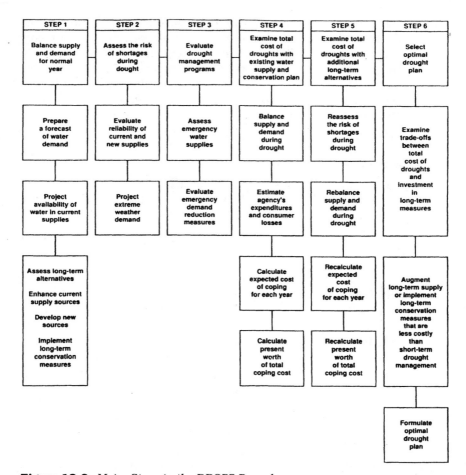

**Figure 12-2.** Major Steps in the DROPS Procedure

during normal supply conditions is assured by adopting plans to enhance the system's current sources of supply and by acquiring and developing new sources, as well as by pursuing conservation strategies for achieving permanent modification of demand. Once the plan for balancing the demand and supply during normal supply years is developed, the selection of the optimal adjustments for drought conditions is based on the assessment of the risk and magnitude of water shortages under the drought conditions.

The trade-off between short-term and long-term adjustments to drought is assessed by estimating the expected present value of the cost of coping with drought emergencies during a prescribed period. This coping cost is determined on the basis of the probability of water shortages and the cost of coping with them. For each year of the planning period, minimum-cost drought response plans are formulated for a range of possible supply deficits. The probability of occurrence of each deficit is assigned to the cost of the corresponding emergency response plan. In this manner, a higher coping cost in any given year is associated with a higher deficit having lower probability of occurrence.

## The Coping-Cost Criterion

The expected value of the total coping cost during a planning period $T$ is found by summing the present worth of the expected values of coping costs in each future year according to the formula:

$$E(TC) = \sum_{t=1}^{T} \sum_{i=1}^{k} p_{it} C_{it} (1 + r)^{-t}$$

in which $E(TC)$ = the present worth of the expected value of the cost of coping with shortages during the planning period $T$; $p_{it}$ = the probability of water shortages in year $t$, which shortage would call for expenditure $C_{it}$ to cope with it; $i$ = discrete levels of probability (e.g., 0.01 for a 100-year drought, 0.10 for a 10-year drought, etc.), $i = 1 \ldots k$; $t$ = year of the planning period; $r$ = discount rate.

The present worth of the expected value of the long-term cost of coping with water shortages is inversely related to the investment expenditure on the long-term supply augmentation and demand-reduction projects. Any long-term alternative will affect the size of future deficits, thus resulting in a different expected value of the cost of coping with droughts. The optimal solution is found at a point where the incremental cost of the long-term adjustments equal the decremental costs to cope with droughts. However, these cost trade-offs do not have the absolute equivalence of monetary values. The relatively "certain" costs of supply augmentation or conservation programs are compared with relatively "uncertain" expenditures and economic losses associated with drought emergencies. In some situations water planners may choose to compensate for the differences in uncertainty by assigning weights to each category, thus possibly accepting long-term protection alternatives which are less cost-effective.

# Example of Phoenix, Arizona

The DROPS procedure was applied while developing a long-term drought plan for Phoenix, Arizona. The City of Phoenix obtains most of its water from the Salt River Project (SRP), an organization formed in 1902 under the name of the Salt River Valley Water Users' Association. The Association pledged about 240,000 acres of land as collateral for a federal loan to finance construction of Roosevelt Dam. Although the debt for construction of Roosevelt Dam and associated reservoirs and irrigation canals was retired in the 1950s, subsequent landowners within the original SRP boundaries have continued to maintain the right to receive water from the system. Hence, water provided to the part of the Phoenix service area located within the original SRP boundaries is called "on-project" water. Historically, about two-thirds of the water received from SRP comes from six reservoirs on the Salt and Verde Rivers. SRP supplements surface supply with groundwater from some 250 deep wells. An additional source of on-project water is groundwater obtained from 10 wells which are operated by the city (Dziegielewski et al. 1992).

In recent years, Phoenix has expanded its water service area to include land outside of the original SRP water rights boundaries. Thus, water provided to this part of Phoenix is known as "off-project" water. The off-project area category as used here also includes some lands located within the SRP boundaries which are not entitled to SRP water rights. The major source of supply for the off-project subarea is Central Arizona Project (CAP). In addition, the city operates 39 wells in this area which can supply water directly to the distribution system and 13 wells which supply water to Salt River Project canals in exchange for an equivalent amount of SRP water. Some off-project water is also obtained from the Verde River as a result of Phoenix building gates in the spillway of Horseshoe Dam operated by the SRP.

## Balance of Supply and Demand

The city has developed plans which balance the supply and demand during a normal supply year and under average weather conditions in the service area (Dziegielewski 1987; City of Phoenix 1986, 1987, 1990). A computerized forecasting model called PHX-MAIN was used to generate estimates of water use in the future based on the projected growth of the service area (Dziegielewski and Boland 1989). The results showed that with the existing long-term demand-management programs, the average demand for water will increase between 1990 and 2040 by 44 percent on-project, and 265 percent off-project. During years with unusually low precipitation and high temperature, as compared to a year with average weather, the projected demand is expected to increase by about 10 percent.

The available water supply will be increased with time to meet the projected demands. In the on-project subarea more SRP water will become available to Phoenix as irrigated agricultural land is converted into urban developments. This supply increase will offset the reduction in groundwater use required by the Groundwater Management Act. In the off-project subarea, additional supplies will be obtained from the expansion of Roosevelt Dam and by the use of reclaimed wastewater for turf irrigation. Additional supply sources will also be developed including imported groundwater from McMullen Valley 80 miles west of Phoenix, additional reclamation and exchange projects, and groundwater recharge and recovery of surplus surface water supplies. Table 12-1 shows the projected demands and supplies during two forecast years: 2000 and 2040.

**Potential Shortages**

The assessment of the risk and magnitude of water shortages under the drought conditions included the consideration of drought conditions in watershed areas of the city's two major surface water sources: the storage system of the Salt River Project on the Salt and Verde rivers and the Central Arizona Project, which draws water from the Colorado River system. The possibility of drought in these watershed areas was assumed to coincide with increased water demands caused by unusually hot and dry weather in the Phoenix water service.

**Table 12-1.** Balance of Supply and Demand in Phoenix

| Demand/Supply (in acre-feet per year) | 2000 | 2040 |
|---|---|---|
| On-Project Subarea | | |
| Demand with existing conservation | 200,100 | 250,300 |
| Above-normal demands w/conservation | 220,600 | 277,400 |
| Normal supplies | 241,400 | 268,100 |
| Supplies during mild drought | 203,100 | 223,000 |
| Supplies during severe drought | 165,100 | 179,000 |
| Deficits | | |
| Under a mild drought | −17,500 | −54,400 |
| Under a severe drought | −55,500 | −98,400 |
| Off-Project Subarea | | |
| Demand with existing conservation | 177,000 | 345,000 |
| Above normal demands w/conservation | 192,600 | 382,900 |
| Normal supplies | 222,000 | 388,900 |
| Supplies during mild drought | 178,200 | 311,500 |
| Supplies during severe drought | 174,600 | 238,900 |
| Deficits | | |
| Under a mild drought | −14,400 | −71,400 |
| Under a severe drought | −18,000 | −144,000 |

1 acre-foot = 1231.7 m$^3$

In order to assess the potential supply deficits during droughts, two drought scenarios were formulated, each designating a different level of severity and probability of occurrence. During a mild drought with a recurrence interval of 10 years (probability 0.10), on-project water supplies may be reduced to about 80 percent of normal deliveries. A mild drought in the off-project subarea would result in the loss of gate-water credits and the added storage at Roosevelt Dam. During a severe 100-year drought (probability 0.01), the deliveries of SRP water to the on-project subarea will be reduced from 3 acre-feet per acre to 2 acre-feet per acre, thus reducing the supply by 33 percent. During a severe drought, the off-project subarea would experience reduced allocation of the CAP water and a complete loss of gatewater and water from new storage in local reservoirs. The Bureau of Reclamation has estimated that the CAP allocation to Phoenix could be reduced from 113,900 acre-feet per year to 45,000 acre-feet per year when a shortage year in the Colorado River system is declared by the Secretary of the Interior. The probabilities of the CAP shortages change over time ranging from 0.01 in 2000 to 0.167 in 2040.

Table 12-1 shows expected deficits that may occur during mild and severe droughts. The growing demand and planned restricted use of groundwater sources will increase the size of shortages in the future. The projected deficits in 2040 for the on-project subarea are expected to range from 54,400 acre-feet per year during a mild drought to as high as 98,400 acre-feet per year during a severe drought (35 percent of projected demand). In the off-project subarea, deficits ranging from 19 to 38 percent can be expected during mild and severe droughts, respectively, at the end of the planning period.

## Options for Dealing with Drought

The options available to the Phoenix Water Services Department for dealing with the projected shortages of water during droughts include (1) additions to long-term supplies for drought protection or permanent modification of demand and (2) formulation of a drought contingency plan to utilize emergency water supplies or achieve temporary reductions in water demand. However, in order to find the optimal balance between the long-term and short-term adjustments, one must determine the cost of coping with drought emergencies given the current and projected balance of supply and demand. Over a long period of time, the total cost of coping with droughts can be relatively small compared to the often very large expenditures needed to achieve long-term modification of demand or to augment normal supplies to meet drought conditions.

## Long-Term Drought Protection Alternatives

The city has 16 inactive wells in the on-project subarea which can be renovated with wellhead treatment systems to remove excessive concentrations of nitrates at the total cost of $10,800,000. The combined production capacity of these wells would be 27,300 acre-feet per year. Seven of the 16 wells with a combined capacity of 15,200 acre-feet per year could be renovated at the cost of $5,000,000. In the off-project subarea, the city has 17 wells requiring nitrate treatment at the cost of $10,800,000 with a combined capacity of 26,900 acre-feet per year. Also seven of the off-project wells with a combined capacity of 12,000 acre-feet per year could be treated at the cost of $5,000,000. The unit operating cost of these wells was estimated at $65/acre-foot, plus $166/acre-foot of an estimated cost of acquiring and recharging CAP water to replace the withdrawals during drought. The annual maintenance cost was assumed the same as for the existing good quality wells (i.e., $31,900 per well per year).

New wells can be developed at a capital cost of $500,000 per a 2-million-gallon-per-day (mgd) well, not including booster pumps or storage facilities. However, new wells drilled in the on-project subarea where groundwater is contaminated with nitrates would require wellhead treatment systems at the cost of $900,000 per well. The annual maintenance cost for the new wells was assumed to be the same as the cost of currently inactive wells or $8000 per well per year. Pumping costs of $50 per acre-foot on-project and $70 per acre-foot off-project were assumed plus the cost of $166 per acre-foot for recharged groundwater.

The city could also enter into interruptible supply contracts with agricultural land owners in the on-project subarea. If farmers would agree to fallow some of their land, the idle acres would be "cut to" Phoenix, making additional water available for municipal use. For the purpose of this analysis, it was assumed that the city would pay farmers a compensation equal to average profits of about $500 per acre. Because 2 acre-feet of water would become available from each idled acre, the cost of this water would be $250/acre-foot. A minimal up-front capital cost of $100,000 was assumed to establish the administrative mechanism for the program. An assumption was made that a maximum of 30 percent of agricultural water rights within the area could be made available for city use in any one year. However, as agricultural land is urbanized, opportunities for this drought protection alternative will diminish.

The city planning staff has identified nine long-term demand management measures that could save up to 47,600 acre-feet per year of water use in the on-project area by 2040. The same nine measures implemented in the off-project subarea would save 81,400 acre-feet per year. Although some of these measures are beneficial even if they do not produce drought protection benefits, their implementation will result in reducing the expected deficits during future droughts. The costs of the contemplated demand-management program range from $5 to $237 per acre-foot of water saved and average $21/acre-foot.

## Analysis of Drought Protection Trade-offs

The computation of the coping costs with and without investment in the long-term drought protection was performed using an electronic spreadsheet. Eleven benchmark years in five-year intervals were selected for the period of analysis extending from 1990 to 2040. For each benchmark year the balance of supply and demand was used in order to calculate the expected deficits during the mild and severe drought scenarios. In the first step, the costs of coping with the probable droughts during each benchmark year were calculated separately for each drought severity. This was accomplished by selecting the least-cost combination of the five short-term drought-response options (i.e., standby wells supply, emergency PI/PE campaign with drought surcharge, sprinkling restrictions, nonresidential conservation campaign and mandatory rationing). If the mandatory rationing program was required to alleviate the deficit, then the other three demand reduction measures were not used.

The costs of the short-term measures were totaled for each drought severity and each benchmark year. These costs were multiplied by probabilities of each drought (i.e., 0.1 and 0.01), then added together and discounted back to 1990 using a 4 percent discount rate. Finally, the annual costs adjusted for risk and discounted were interpolated for intermediate years and totaled for the entire period of analysis. The result represented the total cost of coping with droughts during the 1990–2040 period without any long-term investment in drought protection.

The Drought Investment Protection Tradeoff (DPIT) for each long-term alternative was calculated as a ratio of the reduction in total coping costs resulting from the implementation of the long-term drought protection alternative to the present value of the investment. For example, the investment (of $17.7 million) in wellhead treatment of nitrate-contaminated wells provided the city with 27,300 acre-feet per year of standby groundwater supply that could be used to alleviate supply deficits during drought.

Table 12-2 shows the results of DROPS analysis for seven alternative drought-protection investments for the on-project subarea. Without investment, the drought coping cost was estimated at $25.1 million. By drilling new wells without nitrate treatment as needed during the planning period in order to alleviate the deficits, this cost would be reduced to only $11 million. The ratio of the reduction in coping cost ($14.1 million) to the present value of the cost of new wells and their maintenance of $17.7 million is 1.76. The DPIT of 1.76 indicates that drilling new wells is a worthwhile investment in drought protection. If the new wells would require wellhead treatment to remove nitrates, then the DPIT would be less than 1, making this alternative economically infeasible. Remediation of all existing wells which are contaminated with nitrates and supplementing them with new wells would also be economically infeasible (DPIT = 0.59). However, the renovation of only seven nitrate-contaminated wells combined with drilling of new wells would have benefits exceeding costs. Also, the investment in long-term water conservation combined with drilling of new wells would produce drought protection benefits exceeding cost.

**Table 12-2.** Drought Protection Investment Trade-offs On-Project Subarea

| Long-term drought protection alternative | Investment in long-term drought protection (1990 $) | Total coping cost with investment (1990 $) | Drought protection investment tradeoff |
|---|---|---|---|
| 1. No investment | 0 | 25,100,000 | — |
| 2. Drilling new wells without nitrate treatment | 8,000,000 | 11,000,000 | 1.76 |
| 3. Drilling new wells with nitrate treatment | 17,700,000 | 11,100,000 | .79 |
| 4. Remediation of all nitrate-contaminated wells plus drilling new wells | 16,500,000 | 15,400,000 | .59 |
| 5. Renovation of seven nitrate-contaminated wells plus drilling new wells | 6,800,000 | 13,100,000 | 1.77 |
| 6. Agricultural land-fallowing program | 100,000 | 11,500,000 | 136.0 |
| 7. Long-term conservation program and drilling new wells | 6,100,000 | 7,500,000 | 2.86 |

The best long-term alternative for the on-project subarea is the agricultural land fallowing. The contracts with farmers would result in a sizable reduction in the coping cost at a very modest investment in setting up the program. However, there is a high degree of uncertainty concerning the willingness of farmers to participate in the program.

Table 12-3 shows the drought protection investment trade-offs for four alternatives for the off-project subarea. Because the expected deficits are much higher and their probability is higher as well, the total coping cost without any additional investment was estimated to be $112.6 million. By drilling new wells during the planning period, this cost would be reduced to $25.2 million at the cost of $20.5 million (DPIT = 4.3). The renovation of seven nitrate-contaminated wells combined with drilling of new wells would also be economically feasible, but less cost-effective than drilling only new wells. Investment in long-term conservation combined with drilling of new wells is by far the most effective alternative for drought protection.

Understandably, the DPIT indicators calculated by the DROPS procedure are sensitive to assumptions about drought probabilities, size of expected deficits, costs of demand-reduction measures, discount rates, and many other assumptions. Phoenix, therefore, tested the effects of varying assumptions on the final results. For example, the economic effectiveness of the agricultural land-fallowing programs would be viable even at much higher costs per acre-foot if the frequency of severe drought for on-project was not 1 in 100 years but 1 in 15 years. As one could expect, all investments become more beneficial as the frequency of droughts increases and a higher cost of emergency demand-reduction measures are assumed.

**Table 12-3.** Drought Protection Investment Trade-offs Off-Project Subarea

| Long-Term alternative | Investment in long-term coping protection (1990 $) | Total protection investment (1990 $) | Drought tradeoff |
|---|---|---|---|
| 1. No investment | 0 | 112,600,000 | — |
| 2. Drilling new wells | 20,500,000 | 25,200,000 | 4.3 |
| 3. Renovation of seven nitrate-contaminated wells plus drilling new wells as needed | 22,000,000 | 33,600,000 | 3.6 |
| 4. Long-term conservation program | 5,600,000 | 15,800,000 | 17.3 |

# Summary

Planning for droughts is an important component of urban water demand planning and management. Demand-side alternatives offer a flexible and cost-effective option for reducing the vulnerability of public water supplies to future droughts. A systematic planning for drought protection and response is a necessary component of integrated water resources planning.

# References

City of Phoenix, Water Resources Plan 1987. Staff Report. Water and Wastewater Department (May 1986).

City of Phoenix, Water Conservation Plan 1986. Staff Report. Water and Wastewater Department (July 1987).

City of Phoenix, Phoenix Water Resources Plan—1990. Staff Report. Water and Wastewater Department. Water Conservation and Resources Division.

Dziegielewski, B., D. D. Baumann, and J. J. Boland. Evaluation of Drought Management Measures for Municipal and Industrial Water Supply. *IWR Report 83-C-3.* Institute for Water Resources, U.S. Army Corps of Engineers, Fort Belvoir, VA (Dec. 1983).

Dziegielewski, B. and J. E. Crews. Minimizing the Cost of Coping with Droughts: Springfield, IL. *Journal of Water Resources Planning and Management, ASCE,* 112:4:419 (Oct. 1986).

Dziegielewski, B., D. D. Baumann, and J. J. Boland. Prototypal Application of a Drought Management Optimization Procedure to an Urban Water Supply System. *IWR Report 83-C-4.* Institute for Water Resources, U.S. Army Corps of Engineers, Fort Belvoir, VA (Dec. 1983).

Dziegielewski, B., E. M. Opitz, D. D. Baumann, W. Y. Davis, and C. M. Padin. *Optimal Drought Plans, Volume II: Evaluation of Economic Losses Resulting from Water Supply Shortages.* United States Department of the Interior, Geological Survey Report (Grant 14-08-0001-G-1069). Washington, DC (February 1986).

Dziegielewski, B. Drought Management for the Phoenix Water Supply System. Consultant Report. Planning and Management Consultants, Ltd., Carbondale, IL (Apr. 1987).

Dziegielewski, B. and J. J. Boland. "Forecasting Urban Water Use: The IWR-MAIN Use Model," *Water Resources Bulletin,* vol. 25, no. 1. (February 1989), pp. 101–109.

Dziegielewski, B., W. R. Mee, Jr. and K. R. Larson. "Developing a Long-Term Drought Plan for Phoenix," *Journal of American Water Works Association,* vol. 84, no. 10. 46–51 (Oct. 1992).

Moreau, D. H. and K. W. Little. 1989. *Managing Public Water Supplies During Droughts: Experience in the United States in 1986 and 1988.* University of North Carolina: Water Resources Research Institute, September 1989.

# Index

Accounting perspective, in benefit-cost
  analysis, 268–269
  summary of benefits/costs by, 270 (table)
Accuracy, correlates of, in water use
  forecasts, 83–84
Ad hoc drought response measures, vs.
  drought contingency plans, 332
American Water Works Association
  (AWWA), xiv, 4, 206
Applicability, initial screening test,
  239–240
Avoided cost, 306

Base forecasts, in forecasting under
  uncertainty, 92–93
Benefit-cost analysis:
  accounting for intangibles in, 269–271
  accounting perspectives in, 268–269, 270
    (table)
  benefit-cost measures in, 272–274
  in demand management planning,
    263–274
    customer program costs, 264
    decreased utility revenue, 264
    external costs, 264
    utility program costs, 264
  process of discounting in, 271–272
  use of IWR-MAIN forecasting tool in,
    106–108
Benefit-to-cost ratio, 108, 273–274
Best Management Practices (BMPs),
  253–254, 256 (table), 278
Bivariate models, of water use forecasting:
  per capita requirements methods, 86
  unit use coefficient method, 86–87
Block rates:
  decreasing, 138
  increasing, 138, 165–178
    case study of, change to uniform price,
      191–219

Broadview Water District, as case study of
  increasing-block rates, 169–172
Brown and Caldwell study, of water conservation savings, 54

*Census of Manufacturing,* 50
Central Arizona Project (CAP), 337–341
CES (constant elasticity of substitution)
  production function, 37–41, 48–49
Clean Water Act (1977; formerly Federal
  Water Pollution Control Act), 8, 225
Cobb-Douglas production function, 37–38,
  40–41, 48–49, 57–61
Colorado–Big Thompson Project (C–BT),
  181
Contingency trees, in forecast uncertainty,
  91
Coping-cost criterion, in DROPS planning,
  336, 341

Demand management:
  historical emphasis on supply alterna-
    tives, 6–7
  integrated resource planning (IRP)
    paradigm most closely associated
    with, 304–308
  options for enhancement of drought
    protection, 331
  planning methods, 237–281
    conduct benefit-cost analysis, 263–274
    define implementation conditions,
      258–262
    determine applicability/feasibility,
      239–240
    determine social acceptability, 240–250
    establish program goals, 238–239
    estimate potential water savings,
      250–258
    integrate water conservation into water
      supply plans, 274–280

Demand management (*Cont.*):
program evaluation methods:
data collection, 288–289
design of evaluation studies, 290–291
measuring impact of, 293–298
role of, 283–288
sample selection for, 289–290
standardizing water savings estimates, 299–301
uncertainty in water savings measure-ments, 298–299
validity of studies, 291–292
social criteria for evaluating development in, xi, 196, 211, 240–250
and trends in water/sewer services revenues/expenditures, implications for:
categories of expenditures, 222–231
discussion of, 233–235
relationships in self-financed utilities, 231–232
water conservation measures in, 7–9
Discounted payback period, economic feasibility test, 108
Discounting process, in benefit-cost analysis, 271–272
Distributed resources, 306
DROPS (DRought Optimization ProcedureS) approach, to drought management, 333–336
major steps in, 335 (figure)
Drought contingency plans, vs. ad hoc drought response measures, 332
Drought event, management of, crucial questions to be answered in, 332
Drought Protection Investment Trade-off (DPIT) indicators, 341–343

Econometric demand models, 88
Economic theory:
of consumer demand for water, 56–65
of demand for water as input, 34–47
implications for empirical modeling of industrial water use, 47–52
Electric Power Research Institute, 306
Embedded cost, 143, 144
Empirical estimation, in potential water savings, 257
End-use relationships, in conservation savings methods, 104–106

Engineering approach. *See* Mechanical estimates
Expenditures, water/sewer service, categories of, 222–225
aggregate trends, 225–227
debt, 230–231
household trends, 227–230
Explanation, as first step of forecasting process, 82

Federal Water Pollution Control Act. *See* Clean Water Act (1977)
Forecast bounds, in forecast uncertainty:
arbitrary, 92
confidence intervals, 92
Forecasting, of urban water use:
criteria for forecasting method selection, 81–83
evaluation of methods of, 84–85
bivariate models, 85–87
multivariate models, 87–88
time extrapolation, 85
expressing uncertainty in, 93
need for, 77–78
in financial planning, 80–81
in long-range planning, 78–79
in short-range planning, 80
in supply adequacy evaluation, 79
principles of, 81
tools for:
IWRAPS©, 108–134
IWR-MAIN Water Demand Analysis Software, 95–108
uncertainty in, methods of expressing:
contingency trees, 91
forecast bounds, 92
"safety-factor," 89–90
scenario approaches, 90–91
sensitivity analysis, 91
use of base forecasts in, 92–93

Groundwater systems, factors bearing on susceptibility of, to drought, 331

Harvard Water Program, xiii–xiv

Impact evaluation, 287–288
data collection for, 288–289
Income elasticities, in North America, 67–72 (table)

Integrated resource planning (IRP):
application to urban drought, xi, 329–344
application to water-sector planning, 306
case studies (water sector):
Metropolitan Water District of
Southern California, 321–324, 325
(table)
New York City, 311–314
Seattle, Washington, 316–318, 319
(table)
Southern Nevada Water Authority,
319–321, 322 (table)
Wichita, Kansas, 314–316, 317 (table)
evolution of, 326
as integration of demand-side/supply-
side alternatives, 19–20
most closely associated with demand
management concept, 304–308
role of cost comparisons in, 306, 308
seven basic steps in process of, 309–310
charted, 309 (figure)
vs. traditional utility planning, 18–20,
304, 305 (table)
IWRAPS© (Installation Water Resources
Analysis and Planning System)
water requirements modeling,
relative to:
building types/sizes, 111
climate, 113
conservation/mobilization, 115
construction/demolition, 112
forecast procedures, 123–134
installation/fixed effects, 114–115
installation mission, 113–114
seasonality, 112–113
use at military installations, 108–110
water requirement algorithms:
conservation, 122–123
sectoral allocation, 120–122
summer model, 118–120
water use coefficients/level of activity
adjustments, 115–117
winter model, 117–118
IWR-MAIN (Institute for Water
Resources—Municipal and
Industrial Needs) Water Demand
Analysis Software, 23, 95–108
background/development of, 95–97
benefit/cost analysis methods, 106–108
conservation savings methods, 103–106
econometric demand models, 88

IWR-MAIN Water Demand Analysis
Software (*Cont.*):
forecasting methods:
nonresidential sector, 100–102
other/unaccounted sector, 103
public sector, 102–103
residential sector, 99–100
inputs and outputs, 109 (figure)
major features of, 97–99

Köppen's climatic classification system, 113,
114 (figure)

Least-cost planning, as setting for introduction of demand management, 19,
304, 306
Levelized costs, economic feasibility test, 108
Lifecycle revenue impact (LRI), economic
feasibility test, 108
Los Angeles Department of Water and
Power (LADWP), as case study of
increasing-block rates, 173–178

M&I (municipal and industrial) use:
industrial water use, 33–34
economic theory of demand for water
as input, 34–47
implications for empirical modeling of,
47–52
relevant economic terms, 32
residential water use, 52–56
economic theory of consumer demand
for water, 56–65
empirical estimates of M&I demand
elasticities, 65–66
synonymous with urban water use,
31
Marginal cost, 143, 144
Marginal cost pricing, 148–149
economic argument for, 149–158
increasing-block rates, 165–166
case studies of, 166–178
seasonal rates, 158–165
Mass, Arthur, xiii
Mechanical estimates, in potential water
savings, 250–252
Metropolitan Water District of Southern
California (MWD), 23–24, 96,
166–169, 321–325
Mobilization scenario analysis, in
IWRAPS© modeling, 128–134

Multivariate models, of water use fore-
    casting:
  econometric demand models, 88
  multivariate requirements models, 87–88
Multivariate regression models, in program
    evaluation, 296–298

Negagallons, 306
Negawatts, 306
Net present value (NPV):
  in benefit-cost analysis, 273
  as economic feasibility test, 107–108
Normalization, 125
  coefficient, 125 (table)

Peak-load pricing, principles of, 159–162
Per capita use, myth of, 22–26
Per capita requirements method, 86
Phoenix Water System, case study of
    water rate structure revisions by,
    191–219
PHX-MAIN, computerized water use
    forecasting model, 337
Pinchot, Gifford, 11
Planning approaches:
  integrated resource planning (IRP),
    19–20
  least-cost planning, 19
  traditional supply planning, 18–19
Prediction, as second step of forecasting
    process, 82
Price elasticities, in North America, 67–72
    (table)
Program evaluation, in demand manage-
    ment:
  data collection for, 288–289
  design of evaluation studies, 290–291
  measuring impact of, 293–298
  role of, 283–285
    objectives, 285–286
    process evaluation vs. impact evalua-
      tion, 286–288
  sample selection for, 289–290
  standardizing water savings estimates,
    299–301
  uncertainty in water savings measure-
    ments, 298–299
  validity of studies, 291–292
Process evaluation, 286, 287 (table)
  data collection for, 288

Quest for Pure Water, The (Baker), 1

Ramsey pricing, 156–157
Residential water use, 52–66
  multifamily, 52, 53 (table), 56 (table)
  single-family, 52, 53 (table), 56 (table)

Safe Drinking Water Act (1974), 8, 315
Safety-factor method, in forecast
    uncertainty, 89–90
Salt River Project (SRP), 337–338
Sanitary Revolution, 6
Scenario approach, in forecast uncertainty,
    90–91
Screening tests, of conservation measures:
  applicability, 239–240
  social acceptability, 240–250
  technical feasibility, 239–240
Sensitivity analysis, in forecast uncertainty,
    91, 98
Showerhead retrofit program, 265–266, 267
    (table)
Social acceptability, 141–142, 196, 211
  determination of, in demand manage-
    ment planning, 240–250
Southern Nevada Water Authority, 319–322
Standardization, 125
  coefficient, 126 (table)
Supply-and-demand management,
    practices/strategies of, 307 (table)
Stone-Geary utility function, 57–63

Technical feasibility, initial screening test,
    239–240
Time extrapolation, water use forecasting
    method, 85
Time preference, 271
Total water management, 306
Traditional supply (utility) planning, 18–19
  vs. integrated resource planning, 304, 305
    (table)
Tucson Water Department, as case study of
    increasing-block rates, 172–173

Ultra-low-flush toilet retrofit program, 259,
    265, 266 (figure)
Uniform-rate variable change, 138
Unit use coefficient method, 86–87
Unit water savings, in potential water sav-
    ings estimates, 252–254

Urban drought:
  DROPS planning framework, 333–336,
    341–343
  and coping-cost criterion, 336, 341
  example of Phoenix, Arizona:
    analysis of drought protection trade-
      offs, 341–343
    balance of supply and demand, 337,
      338 (table)
    long-term drought protection alterna-
      tives, 340–341
    potential shortages, 338–339
  four types of adjustment to, 334 (figure)
  long-term protection against, 330–333
    proactive drought management and,
      333
    short-term drought management and,
      332–333
  need for new IRP water management
    strategies against, in relation to:
    competition for funding of major
      construction programs, 330
    limited availability of untapped supply
      sources, 330
    more active public role in resource
      management decisions, 330
    more stringent water purity standards,
      330
    new environmental legislation, 330
Urban water use:
  determinants of:
    industrial water use, 34–52
    residential water use, 52–66
  forecasting:
    models and applications of, 95–134
    theory and principles of, 77–93
  mean per capita, 3 (table), 4
U.S. Army Corps of Engineers, 7
  Institute for Water Resources (IWR),
    xiv, 96
U.S. Bureau of Reclamation, 153
U.S. Bureau of the Census (USBC), 223–224,
  227, 230
U.S. Department of Interior, 10, 96
U.S. Federal Water Resources Council,
  7, 10
U.S. National Water Commission, 7

Washington Public Power Supply System
  (WPPSS), 153

Wastewater:
  disposal, 4
  flow, linked to water use, 4
Water banking, 9
Water conservation:
  alternate proposals, development of:
    combined objectives, 277–278
    economic objective, 276–277
    environmental objective, 277
  debate concerning nature/definition of,
    9–17
  definition of, 16–17
  documentation of plans, 279–280
  initiatives:
    hydrant locking devices, 312
    residential water survey program,
      312
    sonar leak detection, 312
    toilet rebate program, 312
    universal metering, 312
  integration of, into water supply plans,
    274–280
  measurement of effectiveness of, 17–18
  myths concerning, 27–29
  potentially feasible/potentially accept-
    able measures of, 278
  proposal, development principles of:
    interaction effects, 276
    merit order, 274–276
    net beneficial effects, 276
  renewed prominence of, 7–9
Water management:
  early history of, 1–2
  economic importance of urban water
    industry, 4–6
  evolution of urban water use, 2–4
  historical emphasis on supply alterna-
    tives, 6–7
  integration of demand and supply alter-
    natives:
    integrated resource planning (IRP),
      19–20
    least-cost planning, 19
    traditional supply planning, 18–19
  measurement of conservation effective-
    ness, 17–18
  myths, 21–29
  water conservation as subset of practices
    of, 7–18
Water markets, 188